The Romance of Tata Steel

The Romance of
Tata Steel

R.M. LALA

FOREWORD BY RATAN N. TATA

PENGUIN
VIKING
An imprint of Penguin Random House

VIKING

USA | Canada | UK | Ireland | Australia
New Zealand | India | South Africa | China | Singapore

Viking is part of the Penguin Random House group of companies
whose addresses can be found at global.penguinrandomhouse.com

Published by Penguin Random House India Pvt. Ltd
4th Floor, Capital Tower 1, MG Road,
Gurugram 122 002, Haryana, India

Penguin
Random House
India

First published in Viking by Penguin Books India 2007

Text copyright © R. M. Lala 2007
Photographs copyright © Corporate Communications Division and Tata Steel Archives
Jamshedpur 2007

15 14 13 12 11

ISBN 9780670081462

Typeset in *Sabon Roman* by by SŪRYA, New Delhi
Printed at Replika Press Pvt. Ltd, India

www.penguin.co.in

MIX
Paper from
responsible sources
FSC® C016779

This is a legitimate digitally printed version of the book and therefore might not
have certain extra finishing on the cover.

This book is dedicated to the pioneers of India's modern steel industry.

To Jamsetji N. Tata, who envisioned the steel plant as a fountainhead of India's industrial regeneration and worked for it thirty years before the first ingot of steel rolled out from his plant.

To Lord George Hamilton, Secretary of State for India during the reign of Queen Victoria, who understood the vision and passion of Jamsetji Tata and helped him.

To the brave men and women who explored the jungles for iron ore and coking coal and then erected in a forest India's first steel mill with elephants and bears as their neighbours.

To those who ran it under great odds at over 100 per cent capacity.

To those who struggled to turn around the plant in the 1990s to make it a world-class company.

And to those who thereafter dreamt and aspired to make it global, carrying with them, wherever they go, its fundamental values and philosophy.

Jamsetji N. Tata
1839–1904

When you have to give the lead in action, in ideas—a lead which does not fit in with the very climate of opinion, that is true courage, physical or mental or spiritual, call it what you like, and it is this type of courage and vision that Jamsetji Tata showed.

—Jawaharlal Nehru

Jamsetji Tata embarked upon setting up a steel manufacturing plant. It heralded a major shift in Indian business from trading into manufacturing. It was an epoch-making event for the entire nation because it signified our first step towards self-reliance in manufacturing...I will always remember Jamsetji, not just as a visionary industrialist but as a man who helped a nation believe in itself.

—Azim Premji in *For the Love of India: The Life and Times of Jamsetji Tata*

He sought no honour,
he claimed no privilege;
but the advancement of India, and
her myriad peoples was with him an abiding passion.

—*Times of India* Obituary on Jamsetji N. Tata, 20 May 1904

Contents

Part IV: Touching the Lives of People

Part V: Struggle and Triumph

Part VI: Going Global

Foreword

Tata Steel's history is both colourful and eventful. The company's fortunes have, in many ways, been woven into the fabric of the last decades of colonial Raj and the development of a new independent India.

Russi Lala has chronicled, in considerable detail, the 100-year history of Tata Steel, starting with the vision of its founder, Jamsetji N. Tata, followed by leadership provided by his successors—Sir Dorab Tata and J.R.D. Tata, who implemented and expanded Jamsetji's vision.

Russi has captured the 'touch and feel' of events in Tata Steel from its early days: its role in the war effort, its contribution to the economy's development in the early days of India's independence and its more recent transformation into a vibrant modern steel plant—recognized internationally as one of the world's most cost-effective steel manufacturers.

Russi also succeeds in bringing to life the human side of the company in a very readable and cogent manner. He has captured how the strength of the company is embodied in the spirit of its people and their unbelievable will to win.

The book is a valuable and interesting record of the company's evolution over its 100-year history, while at the same time being an enjoyable book to read.

Ratan N. Tata

Preface

It is not important how long a man lives, but how well. It is the same for a company. It is not a centenary of existence that is of prime importance about Tata Steel, nor its tonnage, nor its high quality of steel. Many can excel in tonnage and equal it in quality.

Its romance lies in its birth—the vision of a man of a subjugated and primarily agricultural country, who envisioned setting up a steel plant to spearhead the industrialization of his country. His desire, as a true patriot and visionary, was to make India an advanced and industrialized power in the realm of nations.

The romance lies in the dedication of other men who understood his vision and worked to see it fructify against great odds. Dedicated men exploring through jungles for iron ore making their way in bullock carts and by foot, and in that process even having to brew their tea in soda water.

The significance of Tata Steel, furthermore, lies in the principles it laid down for its operations at the outset, like an eight-hour working day in the steel mill, while in the West it was still a twelve-hour working day. In the decades to follow, it set standards of social welfare which were officially enforced by law for other industries—five, ten, twenty and thirty years later. Over these 100 years it struggled at times, stumbled in its labour relations in the early years, but it learnt from it all and emerged as a company that has not had a strike in sixty-five years. Its labour relations are unique and participative. The company gave unprecedented facilities to its workers, rooted in the practice of Tata's first enterprise, the Empress Mills. Jamsetji Tata said, 'We do not claim to be more unselfish, more generous or more philanthropic than other people. But we think we started on sound and straightforward business principles, considering the interests of the shareholders our own, and the health and welfare of the employees the sure foundation of our prosperity.' These social benefits were further advanced and improved in TISCO's case by getting inputs from two well-known socialists, Sydney and Beatrice Webb, who were invited from England after the First World War.

J.R.D. Tata, chairman of Tata Steel for forty-six years (1938–84) set out his own guiding principles, one of which was:

'Nothing is worth attempting that will not benefit the nation.'

And in that pursuit he laid down certain ethical standards which four successive chairmen of Tata Steel have upheld.

Ratan N. Tata, in his Epilogue to *The Creation of Wealth*, said that he has tried to maintain the business tradition of the Tatas and expressed the hope that 'amidst fast eroding values, [it] will continue to stand out as a well integrated, growth oriented group that marked leadership, operating with higher level of integrity and a great value system and uncompromising goal to achieve results without partaking in corruption, bribery and/or political influence'.

In a review of the same book, the *Commonwealth Lawyer*, published from London, noted, 'The sheer scale of the growth of the Tatas, from a small trading firm in the late-nineteenth century to an industrial behemoth which embraced such diverse areas of activity and products as iron and steel, energy, chemicals, higher education, scientific research, automobiles, hospitality, cosmetics, tea, software development and consultancy, and textiles, coupled with a dazzling record of genuine philanthropy, is awesome by any standards. That all this was achieved without ever departing from a firm and unshakeable commitment to what would today be called "corporate social responsibility", makes it all the more impressive. (A measure of the esteem in which the management of the Tata enterprises was held can be had from the fact that when the Indian Government, in a blatantly populist move, attempted to nationalize the Tata Iron & Steel Company Ltd [TISCO] in the 1980s, one of the most vocal opponents of the move turned out to be the company's trade union, no less!) What emerges from the book is a remarkable story of wealth creation for the public weal—a story which, in our increasingly cynical age, might yet open a few eyes to the highly beneficial potential of ethical capitalism.'

In 1969, after TISCO had done all it could for Jamshedpur and the workers there and in the far-flung mines and collieries, the then chairman, J.R.D. Tata, said that industry should care for the surrounding areas. The community gained substantially and Tata Steel, along with some other companies, amended its Articles of Association, with the shareholders' permission, to go beyond the immediate interests of the company and its staff.

The labour relations it was so proud of could have faced serious strain in the 1990s when, to survive, it had to drastically reduce its 78,000-strong labour force. It was down to 54,000 in 2000 and in 2006 it stood at 38,000. Few companies could have survived such a measure and still maintained what the World Steel Dynamics calls 'a good company culture'.

Not everybody is aware that in the 1990s, Tata Steel, according to a respected international consultant, was heading for possible extinction, struggling as it was with outdated machinery and with government-controlled prices. The struggle for survival is brought out towards the end of the book.

I have singled out at some length the unique measures of the company to care for people even in their personal problems. I witnessed an incident I have mentioned of a broken marriage of a company employee being reconciled and a wealth of instances of non-Tata villagers benefiting in health, education and the like.

During TISCO's struggle for survival in the mid-1990s, the prime minister had called a meeting of the heads of big companies suggesting they give one per cent of Profit After Tax to social work. Calculations revealed that even in its years of low profit Tata Steel had spent a higher percentage on social welfare than expected. The figures ranged from 4 to 13 per cent—13 per cent in its most difficult year of poor profits and 4 per cent in its most prosperous year. The company was created to produce steel, but guided by the vision of its founders, unknowingly, step by step, it was evolving for India the model of a new industrial culture. In the growing circle of management colleges like the Indian Institute of Management (IIM), there are more case studies from Tata Steel than any other company—twenty-one in all to date.

The romance that began 100 years ago has not quite ended. As the company is modernized to be a world-class company it is moving beyond India's boundaries for new ventures. 'Grow old along with me...the best is yet to be'.

1 May 2007   *R.M. Lala*

Part I

A Dream and a Creation

There was only one man who looked boldly to the future and peering beyond the hand-to-mouth practice of his competitors discerned the possibility of an India which would embrace the complete economic cycle. That man was Jamsetji Nusserwanji Tata.

—Sir Stanley Reed
Editor, *Times of India,* 1907–1923

1

Man of Destiny

It was the decade of the 1880s. The dense forests of the Central Provinces teemed with tigers, elephants and other wildlife.

A thin line of travellers wove its way through the jungles. In front, on horseback, was a sharp-featured, bearded young man, behind him on another horse sat a remarkably beautiful woman in a sari, riding side-saddle. Another pony followed with two babies on it in the charge of a competent caretaker. Following at a respectable distance behind was a camel loaded with tents and chattels. Atop the camel were perched the servants of the party.

Every ten or fifteen miles the travellers would stop for the man at the head to spend four or five days on field survey, searching for iron ore, mica, coal and other minerals. He was Pramatha Nath Bose, the first graded Indian officer in the Geological Survey of India. The lady was his wife, Kamala, daughter of Romesh Chunder Dutt of the Indian Civil Service (ICS), litterateur and one of the figures of the Bengal renaissance. When the day was over and the work was done, they would sit round with local tribals, the *Kols* and the *Bhils*, and Bose would invite them to speak about their manners and customs. The Bose family enjoyed the conversations immensely. The people were simple. They earned their livelihood by hunting game, cultivating vegetables and rearing poultry. The party carried tinned food sufficient for six months consumption. The camel, loaned to them by the government, was well-loaded. For six months of the year, from October to March, the family was on continuous safari. Come April, they boarded the train bound for Calcutta and, notes Kamala Bose, 'it was a great pleasure to see Bengali faces and hear them talk Bengali after an absence of six months.'

P.N. Bose, the meticulous geologist who discovered iron and limestone in Chanda and subsequently led the Tatas to Mayurbhanj.

It was P.N. Bose who discovered iron ore in the Chanda and Durg districts of central India. On retirement from service in 1903, he was requested by the far-sighted maharaja of Mayurbhanj to explore the mineral deposits of the state in northern Orissa. This Bose did successfully.

About the same time, as the Boses were braving the jungles of the Central Provinces, on the west coast of India, a highly successful textile magnate was flipping through the pages of a report by Ritter Von Schwartz on 'The Financial Prospects of Iron-making in the Chanda District'. The 43-year old industrialist was Jamsetji Nusserwanji Tata. Almost two decades earlier, as a young man, he had attended a lecture in Manchester by Thomas Carlyle who said that 'The nation that has the steel will have the gold.' That comment had stayed with him since.

India had an ancient iron industry and 'the claims of India to a discovery which has exercised more influence on the arts conducive to civilisation and the manufacturing industry than any other within the whole range of human invention is all together unquestioned,' J.M. Heath said of India in 1839. India's iron deposits had been worked for thousands of years. Great heaps of slag were found in Wai at the foot of Panchgani in western India.

The oldest iron monument in the world is the famous Iron Pillar near the Qutub Minar in New Delhi. Built in about AD 300 it stands without rust, a symbol of the accomplishment of Indian craftsmanship. Another iron pillar is at Mandu, not far from Indore. The Mandu pillar was double the height of the Delhi pillar and though now broken into three pieces, it was probably the largest mass of iron known to the ancient world. Wootz Steel of Hyderabad was the material from which the Damascus blades were furbished. The art of damascening upon soft steel was widely practised in India and the Arab Edrisi noted: 'The Hindus excel in the manufacture of iron...It is impossible to surpass the edge you get from Indian steel.'

In 1878, at the International Exhibition in Paris which Jamsetji visited, there was a remarkable collection of Indian weapons. But all that sophisticated craft seemed to have withered away and what remained were primitive village furnaces. During those days, every ton of steel was imported into India. In 1882, when Jamsetji seriously thought of setting up a steel plant, the world steel production was only 4.2 million tonnes. Today, Tata Steel alone produces more than that. It was the railways which were the first big users of steel, and bridges and ships were to follow.

In 1889, the Eiffel Tower was inaugurated. It focused for the world the use of steel as construction material. For Jamsetji Tata, steel was the archway through which India would enter the age of the industrial revolution.

Von Schwartz's report of 1882 indicated to Jamsetji

Tata that the time had come to start the steel industry in India. He visited Lohara in the Central Provinces and collected samples of iron ore, and from Warora he got coking coal. Though the iron ore was of suitable quality the coking coal was not. Government licensing policy at that time prevented acquisition of mines and raw materials. So Jamsetji did not go ahead with the project. But the dream of starting a steel plant in India abided with him. For the next seventeen years, he maintained a scrapbook of newspaper clippings on Indian minerals.

By the 1890s, Jamsetji had become a very wealthy businessman, and everyone expected that he, who had established two successful textile mills, would start a third one and a fourth one. But by that time something very profound seems to have happened to Jamsetji and he decided that from then on his business was not his main business. The nation was his business. In pursuance of that passion he decided to give India, when he was in his sixties, a steel industry, a hydro-electric project and a scientific institution, the like of which even England could not boast of at that time.

Writing about Jamsetji in his 'Foreword' to *J.N. Tata: A Chronicle of His Life* by Frank Harris, J.R.D. Tata observes: 'That he was a man of destiny is clear. It would seem, indeed, as if the hour of his birth, his life, his talents, his actions, the chain of events which he set in motion or influenced, and the services he rendered to his country and to his people, were all predestined as part of the greater destiny of India.'

It was in the search for that greater destiny that Jamsetji decided in 1892 to set aside Rs 25 lakh to train Indians abroad primarily for admission in the ICS, for the Bar and medicine. In 1896, he unfolded his plan for a university of research, setting aside for it fourteen of his buildings and four landed properties in Bombay, worth about half his fortune. He took as the university's model the Johns Hopkins University in Baltimore. He wanted India to have the technological capability to walk erect into the industrial age. The viceroy, Lord Curzon, could understand neither the greatness of the gift nor that of the giver. It was only after Jamsetji died that Curzon gave the green light for the Indian Institute of Science at Bangalore.

The university finally started in 1911. The only Indian party that had joined in setting it up was the government of Mysore with a generous donation of over 300 acres of its palace grounds plus some financial help.

In 1899, the steel scheme was revived. Major (later General) R.H. Mahon, an Indian Army artillery officer, published a report on the setting up of an iron and steel industry. It recommended Calcutta or its neighbourhood as the best site to make use of coal from Jharia and iron ore from Chanda or Salem with the limestone coming from Burma, at that time administered by the viceroy of India. Jamsetji Tata was intrigued by this report.

Fortunately, the viceroy, Lord Curzon, liberalized the licensing policy on mining, and in 1900 Jamsetji went to England to meet Lord George Hamilton, the secretary of state for India.

He told Hamilton that he had first thought of the steel plant when he was an ambitious young man who wanted to make his way into the world. Now he was over sixty and had made his fortune. If he undertook this venture it would only be for the sake of India. Could he expect cooperation from the government? Hamilton, who had high regard for Jamsetji, was encouraging. Curzon, Hamilton assured him, would surely help. Jamsetji replied, saying, that viceroys come and viceroys go, and before long he may find himself dealing with strangers who would treat him with indifference. He knew his India. Hamilton promised he would authorize the government of India to give him all possible help. Hamilton kept his word. Jamsetji knew not only India but also how British rule operated. He foresaw that a project like a steel plant would need substantial expenditure on the part of the government, like the construction of railway lines and the like, and both the India Office in London and the viceroy needed to work together for it.

After the meeting with Hamilton, in 1900, Jamsetji cabled his office in India to apply for licences to explore in the Chanda district. However, as Jamsetji did not have men with sufficient expertise within India to explore, there was little progress. In 1902, he sailed for England, and then proceeded directly to the USA to study the latest developments in steel technology. On his return to England he saw Hamilton again and observed the coking coal making processes in Birmingham, Alabama, and visited the world's largest iron ore market in Cleveland. He wanted American technology but not the living condition of the workers and staff of their steel cities. While in the USA in 1902, he wrote to his son Dorab:

A copy of the Prospecting Licence granted to the Tatas to look for iron and limestone in Chanda.

'Be sure to lay wide streets planted with shady trees, every other of a quick-growing variety. Be sure that there is plenty of space for lawns and gardens. Reserve large areas for football, hockey and parks. Earmark areas for Hindu temples, Mohammedan mosques and Christian churches.'[1]

In Pittsburgh, he met Julian Kennedy, the foremost metallurgical consultant of the day. Kennedy warned the enthusiastic albeit ageing Indian that even preliminary

[1]For further facts, see the biography of J.N. Tata, *For the Love of India*, by R.M. Lala, Penguin, 2004.

investigations would cost a fortune and there was no guarantee of returns. He would set up a steel plant, said Kennedy, only if a thorough scientific survey was made of raw materials and conditions. He recommended Charles Page Perin in New York as the best man to undertake the geological survey. And so to New York, Jamsetji went.

Perin was poring over some accounts in the office when the door opened and a stranger in a strange garb entered.

He walked in, leaned over his desk, looked at him fully a minute in silence. Finally, he said in a deep voice, 'Are you Charles Page Perin?' 'Yes,' said Perin.

The stranger stared silently for a long time. Then slowly he said, 'I believe I have found the man I have been looking for. Julian Kennedy has written to you that I am going to build a steel plant in India. I want you to come to India with me to find a suitable iron ore, coking coal and the necessary fluxes. I want you

Charles Page Perin, who helped Jamsetji realize his dream of India's first steel plant.

to take charge as my consulting engineer. Mr Kennedy will build the steel plant wherever you advise and I will foot the bill. Will you come to India with me?' Asked later, how he dealt with such a sudden proposal, Perin replied, 'I was dumbfounded, naturally. But you don't know what character and force radiated from Tata's face. And kindliness too.' It was Jamsetji's last business trip abroad.

Before Perin arrived, he sent his partner, C.M. Weld, to prospect for the raw materials. The search had begun in earnest.

2

The Ring of Metal

C.M. Weld was a thin, wiry gentleman who, in his photograph, appears lost in an enormous sola hat. He arrived in April 1903 and in the summer heat of central India set out on his exploration with Jamsetji's elder son, Dorab and a cousin, Shapurji Saklatvala, who was later to be elected to the British House of Commons. Chanda district was one of the finest for hunting. The trouble was they were not hunting for tigers but iron ore. Weld turned out to be a meticulous geologist. The party travelled by bullock cart. There were hardly any roads. Potable water and food were difficult to obtain, and they were often compelled to brew their tea in soda water. At times, they slept in village huts or tents, at other times they slept in the bullock carts for it was not safe to be exposed to wildlife on the ground. As days went by, the immensity of the task they had undertaken began to dawn on the prospectors.

Meanwhile, Curzon, either impatient or convinced that Indians could not deliver the steel, urged Sir Ernest Cassels, a renowned industrialist to prospect for iron ore in Jabalpur.

Sir Ernest sent out two experts from England to explore the area. Jamsetji was not worried; he had already explored it. The two Britishers drew a blank.

Weld found iron ore and limestone in Chanda but the iron ore was in pockets and the district was short of coking coal. Sadly the Chanda scheme had to be abandoned. Weld was all set to go home because he felt that any businessman would try to cut his losses on an expensive consultant. It was only when Jamsetji invited Weld to stay on and explore elsewhere for iron ore, coal and fluxes that Weld said:

'I soon learned that Mr Tata's plan to manufacture iron and steel was inspired by something far broader and deeper than the mere hope of adding to his fortune by a successful iron works in Chanda. His great cotton mills at Nagpur had become an important industrial school for Indians. He strongly felt the need for further industrial outlets for the Indian people. And he dreamt of a successful iron and steel works as another great industrial school, and another step towards that industrialization of his countrymen which he believed was so necessary for their ultimate national prosperity. Therefore, instead of being shipped back to America, I was invited to stay and make an exhaustive study of the possibilities for a commercially successful iron and steel plant in India, regardless of locality. The invitation was too fascinating to be resisted. My whole plan of campaign was thereby altered.'

Weld continues: 'I laboured on all these problems for a year or more, not leaving India for home until July 1904. It would make a very long story if I were to tell all our successive hopes and disappointments, our joys and our tribulations. I travelled many miles by railroad, and many more on horseback through jungle districts with a train of bullock-carts. A part of the time, only elephants would serve for transportation.' 'Jamsetji', he added, 'was always interested, patient and persistent.'[1]

The next breakthrough came from an unexpected quarter. Dorab Tata went to the Nagpur Secretariat to tell the chief commissioner of Chanda that the Tatas had abandoned the prospecting there. As the Commissioner was out, Dorab Tata aimlessly drifted into the museum opposite the Secretariat, to await his return. There he came across a geological map (in colour) of the Central Provinces. Dark colours indicated heavy deposits of iron ore in Durg, 140 miles from Nagpur. These had been indicated by the earlier survey by P.N. Bose. And so to Durg they went.

Competition from home was not lacking. It came not from industrialists but speculators who had got wind that the Tatas were on a treasure hunt at Durg. So they shadowed the prospectors hoping to buy a lease in the neighbouring area so that when the Tatas struck ore they could sell it back to them at a fabulous price.

One fine morning Weld set off in a bullock cart looking very relaxed, armed with a rifle. The shadowers thought he was going game-hunting and decided to take a holiday themselves. Weld reached the village he was looking for and found some iron smelters who worked with primitive furnaces. He asked them where they got their ore, and they took him to a hill about 300 feet high. 'We get it from this hill,' they said. Weld climbed the hill and was astonished to find that his footsteps rang beneath his feet as though he was walking on metal. That was precisely what he was doing. He had found a veritable hill of almost solid iron! In the entire history of the iron industry, never had such a striking and remarkable discovery been made. Not far away, was another hill which was chiefly composed of iron as well.

Hurrying back to Nagpur at once, he applied for a prospecting licence for the Dhalli and Rajhara hills. The

---

[1]*Tisco Review*, November 1933.

licence was granted. The coking coal from Jharia, mentioned in MacMohan's report, and the Dhalli–Rajhara iron ore specimens were hauled to Bombay, and shipped for processing to Germany and America. The reports said they combined well. Now that the raw material had been found in considerable quantity, the next question was where would the plant be located? A suitable site for a steel plant needed abundant water all year round. Weld detected a possible site on the Mahanadi river, a few miles west of Sambalpur near a village called Padampur where there was both limestone and abundant water. It lay mid-way between the iron ore deposits of Dhalli–Rajhara and the coal fields of Jharia. Just as they pitched their tents and drew up the plans to study the course of the Mahanadi river and the viability of the location, news arrived that Jamsetji had died in Germany in May 1904.[2]

Meanwhile, the hand of fortune intervened again. A letter arrived from P.N. Bose towards end-February at the Tata office in Bombay informing the Tatas about the finds of iron ore in Mayurbhanj state, within easy range of Jharia and Raniganj.

On 24 February 1904, P.N. Bose had written: 'I have to bring to your notice an exceedingly rich and extensive deposit of iron ore which I have just explored in this state. The ores consist of magnetite, haematite and limonite. They occur in such abundance that for all practical purpose, they may be considered to be inexhaustible, and limestone of good quality occurs close to them. Almost touching the ground where the iron ores occur, there is a considerable area where the alluvium is more or less auriferous. Among minor minerals, I have come upon asbestos, opal and agate. Altogether, the area is one of great mineral possibilities. Should you entertain the idea of starting iron works in their

---

[2]The work of the Tata expedition was not wasted. Fifty years later, the Dhalli–Rajhara deposits were utilized for the government's steel plant at Bhilai.

state, His Highness the Maharaja will, I have no doubt, afford you every facility and grant you liberal concessions.'

Through all this, Jamsetji's health deteriorated. He was taken to Germany where he passed away on 19 May 1904. Despite the sorrow of his passing, work continued keeping both Jamsetji's memory and his legacy alive. Destiny beckoned the Tatas further east, nearer to Calcutta, to the state of Mayurbhanj in Orissa. Once things had settled down, Weld immediately sent Srinivas Rao, his assistant, to investigate the report and the latter confirmed that there were, indeed, large quantities of rich iron ore in Mayurbhanj.

Soon afterwards, Weld, accompanied by P.N. Bose and Perin this time, experienced once again the dangers, discomforts and exhilaration of exploring a trackless jungle. They were guided on their way by a succession of tribal iron workers who, for centuries, had exploited the deposits of ore in their humble furnaces. At last they found, in the lofty Gurumahisani Hill, rising to nearly 3000 feet, a superb store of iron ore, later estimated to amount to some thirty-five million tonnes; it was mainly haematite with an average iron content of over 60 per cent. It was not quite so extensive or quite as rich as that in Dhalli–Rajhara, but it was much more favourably situated, as the hundreds of acres of rich 'float-ore' lay loose on the surface, where it could simply be picked up by unskilled labour. Indeed, until the end of 1928, the whole of the ore collected at Gurumahisani was float-mixed with earth on the plains and flanks of the hill. The explorers had found a treasure house 'far more potentially valuable than most gold mines'. There were also other hills, almost equally rich, that were mined in later years.

Jamsetji Nusserwanji Tata

Sir Dorabji Tata, Jamsetji's elder son, who played a crucial role in realizing Jamsetji's dreams after his death.

J.R.D. Tata took over as chairman of the Tata group in 1938. Over the next fifty years, under his leadership, the company grew into a business conglomerate of international acclaim.

Ratan N. Tata took over as chairman of the group in 1991. He is responsible for the hugely successful merger with the Anglo-Dutch steel giant, Corus that took place in 2006.

The discovery of the Kharkai and Subarnarekha confluence ended the search for a perfect site for the Tata steel plant. At last, along with iron, coal and limestone, there was plenty of water available.

Dimna Lake, located 13 kms from the city, at the foot of the Dalma Hills.

3

Hidden Wealth Emerges and the Discovery of Sakchi

After the acceptance of Mayurbhanj with Gurumahisani and the surrounding areas of iron ore as the base for their mining operations, the Tatas started looking for a site with suitable railway connections and proximity to the Bengal coalfields. It was decided that the railway station and town of Sini on the Bengal–Nagpur railway would be the site of the new steel works and the railway was prepared to build a line from Mayurbhanj to carry the ore to the small junction called Sini. Having secured these facilities of iron ore and rail transport, the Tatas began the search for finance.

The question that confronted the Tatas now was: 'where was the money going to come from?' They had already sunk about Rs 5 lakh in their explorations over the years, which, for those days, was a lot of money. Everyone in Indian financial circles advised them that the large capital required for a steel plant was just not available in India and that they needed to tap the financial market of London.

Unfortunately, the London market was passing through a bad patch at the time and financiers in London, unsure of the ability of Indians to successfully set up a steel plant, wanted to exert control if they were to invest. After much deliberation, the Tatas decided to take the plunge into the Indian market and issued their prospectus in 1907.

There was a certain debate over the name of the company. Some of the family were of the opinion that *Bapuji* (Jamsetji) did not like the association of his name with his enterprises. Neither for his mills nor for the Indian Institute of Science had he allowed the use of his surname. They forgot that over a decade earlier Jamsetji had enthusiastically started a Tata Shipping Line which had been driven out of business by the P&O led foreign shipping lines repeatedly lowering their prices.[1]

The family's suggestion was overruled. There was a clamour from outsiders: 'Only confidence in the name of Tatas will bring in the capital.' And so the company was registered on 26 August 1907 as the Tata Iron and Steel Company Limited, later referred to as TISCO. Subscriptions opened from early morning till late at night. People besieged the Tata offices in Bombay, 'old and young, rich and poor, men and women they came offering their mites', and within three weeks 8,000 investors had subscribed. Later, when debentures were issued to provide the working capital, the entire issue of £400,000 was subscribed by the maharaja of Gwalior. The Tatas had raised Rs 1.5 crore in ordinary shares, Rs 75 lakh in preference shares, and Rs 7 lakh, in deferred shares. In all a total of Rs 2.32 crore had been raised .

Dorab Tata wrote proudly: 'For the first time in India's financial history, I had succeeded in raising for industrial purposes such a vast sum from the hidden wealth of India for the development of our mineral resources. It was the first time that the raw materials of India did not go out and

---

[1] See the author's book on Jamsetji Tata referred to earlier.

return as finished articles to be sold in the country. Above all, it was a purely *svadeshi* enterprise, financed by *svadeshi* money and managed by *svadeshi* brains.' The timing of the Tatas moving into the financial market is significant. India was just recovering from the feeling of helplessness that had pervaded since the revolt of 1857, and the *svadeshi* movement was giving India back her confidence.

Jawaharlal Nehru was to write years later: 'From 1907 onwards for several years, India was seething with unrest and trouble. For the first time since the revolt of 1857, India was showing fight and not submitting tamely to foreign rule.' In 1904, Indian nationalism had received a shot in the arm with the war victory of Japan, the first Asiatic power to get the better of a European nation, Russia.

It was Jamsetji's company, Tata & Sons, formed 1887,[2] which had obtained concessions for iron ore, rail-freight, and had taken the risk and burden of the exploration. All the concessions were turned over to TISCO for an allotment of shares worth Rs 15 lakh in the new company and out-of-pocket expenses of Rs 5,25,000 to be reimbursed in cash. This Tata & Sons put in equity, adding Rs 4,75,000 of their own money. The overall Tata stake was Rs 25 lakh—about 11 per cent of the total capital subscribed. The TISCO prospectus had announced that the works would be situated at the railway station of Sini but fortune had another surprise in store even after the money was raised.

C.M. Weld, the man who had done much of the exploration for the Tatas, was summoned once again from the USA, after the company was formed. When he arrived at Sini at the end of November 1907, Weld was greeted with the news that the proposed reservoir at Sini, which they were counting on, had proved impractical. The place

---

[2]With Dorab Tata and Jamsetji's cousin R.D. Tata as partners, Jamsetji's younger son Ratan joined later, and in 1917 the company's name was changed to Tata Sons.

had to be abandoned for lack of water. It was a crisis point. Weld and his assistants immediately started to look for an alternative site in the locality which had the same advantages as Sini, provided water could be found. He himself recalls how:

'Early one morning in December 1907, Srinivas Rao and I mounted our horses at Sini and took to the river bed where the Subarnarekha is crossed by the Bengal–Nagpur Railway branch to the coal fields. For some hours, we plodded downstream, our horses making heavy going through the sand. At length, we came upon a sight which filled us with joy: a black trap-dike, crossing the river diagonally, and making an almost perfect natural pick-up weir. It seemed too good to be true.'

Excitedly, Weld and Srinivas clambered up the tongue of land sticking out at the meeting point of the two rivers. When Cortes discovered the Pacific, he stood 'silent upon a peak in Darian'. But Srinivas and Weld shouted with excitement as Archimedes had, when he jumped out of his bath. They looked around the area and found themselves close to the village of Sakchi. They were at the meeting point of the two rivers, Kharkai and Subarnarekha, which together never run dry. There was even a railway station called Kalimati a few miles away. Their long search was at an end.

Weld telegraphed the Bombay office that a better site than Sini had been located. At the Tata headquarters in Bombay, there was speedy acceptance of Weld's findings. A lease was obtained from the Dhalboom Syndicate, over 20 square miles for Rs 30,000, and the simplest of documents drawn up by McNair of Morgans and approved by E.C.B. Acworth of Little and Company, Bombay. The works were to be established at Sakchi—later to be called Jamshedpur.

When the first team of construction engineers and other experts arrived at Sakchi, they were taken in a triumphant procession of tongas and bullock carts with Alexander

Sahlin riding on an elephant as senior partner of the consulting firm of Kennedy, Sahlin & Co. Immediate plans were made for a concrete pick-up dam and the purchase of equipment and plant.

The first party arrives at Sakchi in 1908.

There were about a dozen little huts in the dreamy village of Sakchi and a visitor noted that 'cock-fighting seemed to be the dominant industry'. One can presume that the inhabitants could hardly comprehend the reason for all this excitement and not long afterwards, decided that they would find more peace on the other side of the river.

Ten weeks after the discovery of Sakchi on 27 February 1908, the first stake was driven into the forest-covered plateau. The man who drove the first stake into the ground little knew that with it, India was staking its own future as an industrial power.

One of the earlier arrivals at Sakchi was John Keenan, who rose to be the general manager.

John Keenan, one of the earlier arrivals at Sakchi.

He recalled that local tigers, enraged at the destruction of their forest home, killed two of the labourers, and an elephant driven frantic by the disturbance of the drills, smashed to smithereens a number of huts near the dam. The superintendent of Railway Traffic found that a she-bear had entered his hut at night to deliver a cub under the table. Keenan recalls how a pregnant Santhal girl in the brick department went into labour while carrying a load of bricks on her head and gave birth in the checker chamber. 'What is more, when the baby was born, she picked it up and walked off to her home with it.'

It took a bit more to deliver an integrated steel plant and a city built around it. Mrs B.J.M. Cursetjee, who joined the Tatas in 1905, recalled the events with telegraphic terseness:

'Watchful Weld returned to the States. Vigorous, forceful, impressive Perin came out, handled successfully negotiations with Government for acquisition of land, freight concessions, Guarantee for the purchase of rails. Julian Kennedy Sahlin and Co. of Pittsburg and Brussels were appointed construction engineers. Out came Mr Axel Sahlin, tall, broad-shouldered, active, no minutes wasted on idle longwinded words, or legal phraseology. Julian Kennedy in the States was the engineering genius. Renkin was sent out as their resident engineer. Blueprints poured in by the score - prints with every screw, nut and joint marked. Old K R Godbole, retired from the P.W.D. was appointed in charge of civil works—roads, buildings, the cooling tank bund. A clean, honest, sound, engineer. What a struggle between blustering, bullying Renkin who wanted work pushed at any cost and patient, white-turbaned, methodical, Godbole. How often [Burjorji] Padshah intervened. All in vain. At last Renkin was recalled. Weld came from the States as the first Works Manager.'

Though this enterprise was managed entirely by Indians, it included a blast furnace crew from Germany (who, incidentally, did not live up to expectations), Americans who came as consulting engineers and other covenanted officers and some Britishers.

The first blast furnace was blown on 2 December 1911, and the first ingot of steel to roll in perfect shape was on 16 February 1912. War clouds were gathering over Europe and thirty months later, the guns of August boomed. Suddenly, this 'spot of reclaimed jungle wilderness (became) a place of international importance. For the Allied Forces, it offered the only potential supply of iron and steel east of the Suez Canal' noted a new arrival, Lillian Luker Ashby, wife of the deputy superintendent of police, Robert Ashby. She arrived at Kalimati station from Calcutta in December 1914. It was 'a joyous reunion' with her husband she writes in her recollections of fifty years called *My India*.[3] She and her two children, Hazel ('Cuckoo') and Vivian, transferred to a coolie-powered rail trolley to go on the Tatas' private railway to the steel works.

'Look, there in the distance,' Robert pointed ahead, 'that glare in the sky from the blast furnaces, that thumping and steady roar—that's the Tatas!'

'Oh, see those flames shoot up! Do they always go that high?' Vivian asked. He wanted to know all the details. His torrent of questions was stemmed only when Robert promised to take him over the entire works before the day was done.

'Is there always so much noise, Daddy?' Hazel asked. 'During every one of the twenty-four hours, Cuckoo. They're trying to finish the other furnaces and mills in a hurry so that they can supply steel and ammunition to the armies.' 'And this big company is owned and operated by an Indian firm, Robert?' I (Mrs Ashby) asked, a trifle incredulously. 'Yes,' he replied. 'The Tatas are Parsees from Bombay.'

---

[3]Little, Brown & Co., Boston, 1937.

'Then those skyward-shooting flames are the beacon lights of a new India, an India with initiative. It will be a different land from that we have known all our lives.'

The new chimney for the boiler plant dated 1918.

4

Creating a City

'Be sure to lay wide streets planted with shady trees, every other of a quick-growing variety. Be sure that there is plenty of space for lawns and gardens. Reserve large areas for football, hockey and parks. Earmark areas for Hindu temples, Mohammedan mosques and Christian churches.'

—Jamsetji Tata in a letter written to
his son Dorab in 1902

When Jamsetji penned these words in 1902, he had already embarked on his mission to create a steel plant in India and during that year, he visited Britain and the United States for the same purpose. In the United States, he met the captains of the steel industry at receptions and dinners held in his honour. When he envisaged his city of steel, not only had the site not been found, but even the explorations had not begun in earnest. Yet he had dreamt it all in his mind's eye. When the land of Sakchi was acquired as the site in 1907, some of his dreams were to be translated into reality though he did not live to see it happen.

The Tatas had the twin tasks not only of building a steel plant in a country where the required infrastructure of technical men and machinery was unavailable, but also of planning and raising a city. The site was marked out for the works and a massive job of clearance was undertaken with bullock carts as the main vehicle of transport. Labour was difficult to find and, as an incentive, those who cut trees, were allowed to take it and the Bengal–Nagpur railway offered wagons for transporting the wood. An old timer called Das who visited Sakchi in about 1908 wrote:

Bullock carts and tongas carrying materials
to the construction site.

'There were no roads from the station to the Sakchi camp. Job seekers had to find their way with much difficulty along the Susingeria jungle (the place where Susingeria gate of the Steel Works stands now). In the beginning, a few tents and thatched huts, dotted amidst a jungle clearance, housed the small colony of people who were helping to lay the foundation of the Steel Works. The pioneers spent a hard and adventurous life, defying all dangers and discomforts. No one could move out of his tent at night—wild animals from the neighbouring hills howled and prowled all around. The normal amenities of life were unknown. A cup of tea or milk was a luxury to us. Water was supplied by *bhistis* who brought it in wooden barrels from the Subarnarekha and distributed it on a rationed basis. Sometimes, we had to go without water for hours, and there were occasions when we had to boil eggs and potatoes in aerated water!'

Three years later in February 1911, Lovat Fraser,[1] an English journalist, noted:

'I walked through street after street of commodious one-storey brick houses, all well ventilated, all supplied with running water and lit by electric lights. Many of the houses had electric fans.'

An entire city was being laid out, the first planned modern city of India. New Delhi and Chandigarh were to follow. A night school was started for employees who wished to improve their prospects. A recreational facility for senior staff for bowling, cricket and other sports was built. Every Saturday, the British and American officers organized racing in the heart of Sakchi, near Karim Talkies. Later, they shifted to the outskirts where the airport is today. Soon they developed an enthusiastic clientele of staff and workers who started unofficial betting. When the directors discovered this, they put a stop to racing. To give the citizens a touch of social life, a local brass band was raised. A makeshift hospital was organized in two sheds. It had five beds and a male nurse. The dream was slowly but surely becoming a reality.

There was a hazard even in arriving at Sakchi and you could be waylaid walking from the station or even taking one of the eight tongas which Pathans drove to the city. Then, as now, trains arrived at odd hours and people preferred to sleep at Kalimati Station till dawn, before they resumed their journey. Each arrival was a great event. As

---

[1]'Iron and Steel in India', *Life of Jamsetji N Tata*, Lovat Fraser, Bombay, Times Press, 1919.

many as ten men would accompany a friend safely to his train or collect him on arrival. People still recall in Jamshedpur how their fathers or grandfathers were pressed into service willy-nilly. Recruitment officers, it is said, occasionally craned their necks into crowded compartments and invited people to stop off at Kalimati, if they wanted jobs as carpenters, welders or in semi-skilled trades. 'You could not afford to be idle in those days,' recalls P.K. Chatterjee, who arrived as late as 1921. 'If a cluster of idle men were located, Company officers who usually traveled on horseback would ask them to come along, give them a brass token and they started working that day or the next.'

The early pioneers (from left to right) Mrs Sahlin,
Mr Alexander Sahlin, Mr Kennedy, Mrs Wells (on elephant)
and Mrs Kennedy (in doorway). January 1912.

Putting up the works itself was a mammoth venture. Most of the main machinery arrived by train from Bombay or Calcutta. Alexander Sahlin arrived from Brussels in 1908 and R.G. Wells from America in 1909. The relentless energy of these two was the driving force behind the works coming up within three years—the blooming mills, the finishing mills, the saw mills, the blast furnaces, the coke ovens, the bar mills, not to forget the power house. The capacity of the two blast furnaces was to produce 1,60,000 tonnes of pig iron and 1,00,000 tonnes of steel. Among the officers were 175 foreigners and the total number of operating crew was 2,000. Including unskilled labour, the strength of the company was 8,500 at Sakchi, while another 10,000 were gathering iron ore at Gurumahisani.

The crew of the steel works and its superintendent were Germans, while the English worked in the Ring Rolling Mill. The clerical staff was chiefly composed of Bengalis and Parsis, with 'a few extremely efficient Parsees in various mechanical departments', writes Lovat Fraser. He continues, 'there were a certain number of Austrians, Italians and Swiss, whilst Chinese worked as carpenters and in the Pattern Shop. This medley of nationalities did not always work very well together.'

Fraser recalls an athletic meet where a fierce quarrel broke out in a tug-of-war between the Americans and the Germans. The Americans complained that just as they were winning, the Germans, taking advantage of the darkness, brought extra men upon their rope in order to obtain a victory! To settle this quarrel, Sir Dorab Tata who was acting as the referee, had to intervene. He solemnly declared that the event should be settled next day in daylight. The Americans won easily.

The management had decided right from the start that everyone would have an eight-hour workday in India, in view of the hot and humid weather. This, no doubt, helped.

Soon, Sakchi became known and the first VIP to arrive was the secretary of state for India, Lord Crewe, in 1912. He was then visiting India with King George V. Sir Dorab laid it on in style for him and arranged a special train from Calcutta with breakfast served. The *Times of India*, a bastion of the establishment, was shaken that a British secretary of state had visited a private Indian company. In its report, it noted that Lord Crewe who, 'since his arrival as Minister in Attendance on the King, has received no deputations and made no speeches...made a slight departure from his practice' and accepted an invitation from Tata & Sons to visit their iron and steel works at Sakchi.

With the start of the First World War in 1914, Sakchi became an arsenal of the war effort. Perhaps this explained why the secretary of state had taken the trouble.

Fresh problems came with the war. Out of the blast furnace crew of thirty-one, two-thirds were Indians, but one-third were German technicians. The Germans were interned as prisoners of war and the Americans stepped in their place. No fewer than twenty-six vessels carrying materials ordered by Tata Steel were sunk through submarine warfare and the war effort took almost 80 per cent of the company's production. The entire eastern theatre of the war from Egypt and Mesopotamia to east Africa had a sole supplier of steel. Not a single ton of the Tata's shell-steel

Rail straightening in the Finishing Department (1911).
The railways were the chief consumers of the steel produced by the Tatas.

was rejected, although they had no special apparatus for making steel for shells. The company sold steel to the government at one-third, even one-fourth, the price fetched in the ordinary market and abandoned making highly profitable ferro-manganese in its blast furnaces (for export to the USA), to convert pig iron into steel required for the war effort. It supplied 3,00,000 tonnes of steel in four years of war and the British acknowledged that the allied victories in Mesopotamia would not have been possible but for the 1,500 miles of railway lines supplied by the company.

The great occasion at the end of the war which Sakchi celebrated was the arrival of Lord Chelmsford, the viceroy.

The white viceregal train steamed into Kalimati station and mounted volunteers from the works escorted the viceroy into town. The viceroy stayed at the new directors' bungalow. Standing on the steps of the directors' bungalow before a large gathering, Lord Chelmsford said:

> 'I can hardly imagine what we should have done during these four years (of war) if the Tata Company had not been able to give us steel rail which have been provided for us, not only for Mesopotamia but for Egypt, Palestine and East Africa, and I have come to express my thanks...It is hard to imagine that 10 years ago, this place was scrub and jungle; and here, we have now, this place set up with all its foundries and its workshops and its population of 40,000 to 50,000 people. This great enterprise has been due to the prescience, imagination and genius of the late Mr Jamsetji Tata...This place will see a change in its name and will no longer be known as Sakchi, but will be identified with the name of its founder, bearing down through the ages the name of the late Mr Jamsetji Tata. Hereafter, this place will be known by the name of "JAMSHEDPUR" (applause).'

So saying, he unveiled on the top of the archway of the directors' bungalow, the words painted in large type:

JAMSHEDPUR. The name of the railway station was changed from Kalimati to Tatanagar.

Jamshedpur became a crucible of India at a time when migration within India was not so common. Pathans from the frontier could take the heat of the blast furnaces; Sikhs became the backbone of the steel-rolling shops; clerks came from southern India and Bengal. Aborigine women carried baskets of coal and other material.

Lord Chelmsford at Sakchi in 1919.

As the works took shape and the town evolved, so did its markets and the social life of its population. Modern shops grew along the walkways. A Kashmiri opened a lace shop; a Chinese, shoe shop; a Japanese, a hairdressing saloon and sure enough, a Gujarati businessman called Narbheram installed himself as the largest trader, selling everything from imported Dodge and Ford cars to apples from Australia. As per Jamsetji's dictates, a mosque, a Hindu temple and a church all came up in due time. Wide roads and spacious bungalows made Jamshedpur a prized city.

The growth of the works spurred other industrial activity as well. A host of new manufacturing companies blossomed around the steel works; a tinplate company, the Enfield Cable Company, the Peninsular Locomotive Company, an agricultural implements company and many others, while a small grey statue of Jamsetji Tata at the works gate smiled benevolently beneath the wreaths of marigold and other fresh flowers, bestowed daily in tribute to him by a grateful people.

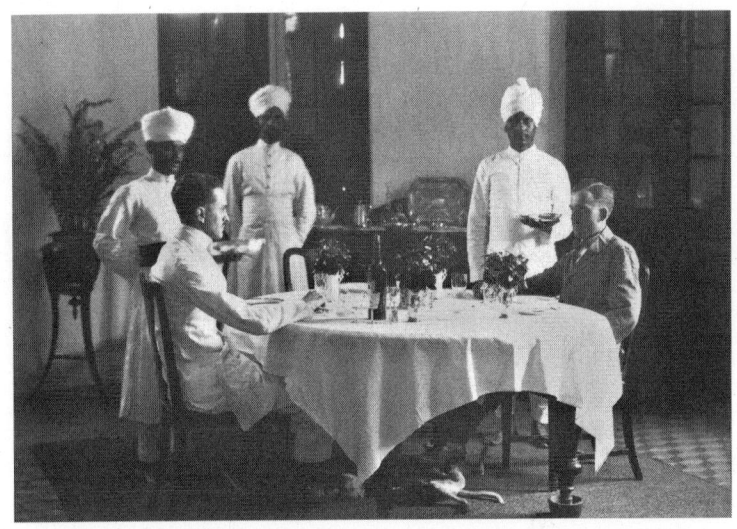

As the steel plant grew so did the city of Jamshedpur and its social life and comforts. Seen here are Robert Sahlin and Lansing Hoyt being served by Rajab Ali Shah, Niamat Khan and Jamil.

Jamshedpur was not only an industrial centre but a destination for many involved in India's struggle for independence. Gandhi, Dr Rajendra Prasad, Jawaharlal Nehru and Subhas Chandra Bose were among the leaders who visited the steel city. Their relationship with Tata Steel and the crucial role they played in its success is discussed in greater detail later.

Mrs Lillian Ashby, wife of the British superintendent of police, recalls Mahatma Gandhi's visit in 1924, at the request of the workers of the steel plant and at the invitation of the directors. On arrival, Gandhi asked Mrs Ashby to show him to his room, in the directors' bungalow. He had hardly entered the room when he objected to the British goods in it, mattresses, pillows, drapes and carpets. Mrs Ashby had it all removed and a handful of Gandhi's personal belongings took their place. 'Will you be quite comfortable with so many things removed from the room, Mahatmaji?' asked Lillian Ashby.

'I am comfortable with just a garment,' was the rejoinder. There, at the heart of India's heavy industry, Gandhi installed his wooden charka.

Mrs Ashby notes that when Gandhi entered Jamshedpur, there were cheering crowds crying out: 'Mahatma Gandhi ki jai.' Flowers and buntings decorated the streets. Years before independence came, she, the wife of a British police officer, wrote of Gandhi's departure:

'That night the Mahatma and his party were driven to Tatanagar station. The lonely outbound path from Jerusalem to Gethsemane was strewn with the fading flowers and dying palms of the morning's wild acclaim. When Gandhi left Jamshedpur, the floral decorations of the streets had withered. There were no cheering crowds, no flaming torches in the hands of enthusiasts. There were only a pitifully few, half-hearted cries, "Mahatma Gandhi ki jai" from the sparse crowd at the railway platform.'

But, she adds: 'The saint had passed that way.'

Part II

Pioneering a New Way

5

A Vision for India, 1916

We are like men building a wall against the sea. It would be the height of folly on our part to give away any part of the cement that is required to make the wall secure. That is why we and you have to use this money...to build up this great industry.

—R.D. Tata, 4 June 1925

Success is a heady drink and the company had more than its share of it. After four years of functioning, in 1916, the chairman, Sir Dorab Tata spoke of:

'Bumper earnings; production 30% above originally designed; cost lower than before...ready and willing markets; capable of yet more expansion...order book full to bursting...'

The company had long-term contracts with the railways and the government was giving every support because of the company meeting their requirements at different points of the war. Tata Steel, in turn, injected confidence in the nation. Though Jamsetji had died twelve years earlier, his vision continued to inspire his successors. Sir Dorab reminded a meeting of the directors at Christmas, 1916, of Jamsetji's vision of a self-supporting, strong, industrial India, of Indian ships carrying Indian goods to far-flung markets across the world. A couple of years later, Sir Dorab articulated this vision in his own words:

> 'So far we have but scratched the surface of this great land. Is it not time that all of us should spare no pains to develop our boundless resources; produce out of the fullness of our own land, and with our own labours, the one thousand and one articles we need for our every day existence and safety? Is it not time to stir ourselves, to wake up to the great opportunity...the task of India's industrial regeneration?'

*Sir Dorab Tata with his wife, Meherbai, Charles Perin
and other members of the management of TISCO
and their spouses.*

This was the motive power behind Tata's ambitious expansion of TISCO even as the war was at its height. At first Perin gave a modest expansion proposal at the board meeting of 1916. Sir Dorab proposed a vote of absolute confidence in Perin and then turning to Perin, Sir Dorab gravely said that he would get from the directors whatever he needed to expand the plant so as to eventually supply India's total steel requirements (of half-a-million tonnes).

There was a moment of silence as the men around the table realized the sweeping significance of the pledge. This pledge made Perin discard his cautious proposals and with them he also discarded his conservative self. And in his place an audacious and ambitious Perin emerged.

The rated capacity of the old plant was 72,000 tonnes of finished steel but it steadily produced over a 1,00,000 tonnes. The Greater Extension Programme envisaged raising the capacity of the plant by five times. To do this, new blast furnaces, coke ovens, machine shops and foundries had to be constructed. Plans were also set afoot to devise the means of rolling steel into various shapes. This was no ordinary venture the Tatas were entering. In order to produce five times the steel they had to simultaneously expand Jamshedpur and its facilities like water supply, housing, sanitation and schools. To feed the new giant that was going to emerge, new collieries had to be taken on lease and in turn townships built around them with the facilities of civilized life.

At the heart of this programme were two men diametrically opposite to each other in character. One was the hard-headed engineer, Charles Page Perin, whose New York office Jamsetji had walked into, in 1902. The other was one of Jamsetji's hand-picked assistants, Burjorji Padshah. Padshah was an idealist and a strict vegetarian who wore canvas shoes. When the maharaja of Mayurbhanj sent a horse carriage to receive him at the station, he declined to use it as the animal had no way of telling him whether or not it was pleased to perform the service for him! Some years earlier a scholarly London club had named seven men 'who really understood Einstein's theory of relativity, men who could use calculus much as ordinary people use subtraction and addition'. Padshah was one of the seven. Had Padshah confined himself to mathematics, Perin may have got along with him. However, Padshah's eccentricities would often confound Perin. In the middle of serious technical discussions, Padshah would open the drawer of his desk, take out a book of poetry and start reading from it. After the initial strain, the one thing that made it possible for them to get on together was that both had unbounded enthusiasm. Slowly they began to appreciate each other and work as a team. They would discuss a small point in the morning, discard it for something better at lunch, mull over it in the afternoon, dream about it at night and finalize it the next morning.

*Burjorji Padshah,
the eccentric genius who
worked with Charles Perin
to implement the Great
Expansion Programme.*

Soon Perin had to return to New York where he hired 300 engineers and draftsmen and, between 1917 and 1920, sent 7,00,000 tracings and 3 million blueprints to India. Despite the 12,000 miles separating them, Perin and Padshah were united by their common purpose, and different as they were, they settled major questions, which would otherwise have taken months, within a matter of days. For example, Perin would cable Padshah saying that he was adopting the new Duplex technique which requires a good deal of pig iron. Padshah would cable back: 'Where are you going to get the molten pig iron for your new process?'

The answer: 'I am designing two new 500-tonne blast furnaces.'

Padshah: 'Where are you going to get the coke for new blast furnaces?'

Perin would wire back: 'I am calling for tenders for three new batteries of by-product coke ovens.'

'Where do you expect to get suitable coking coal?'

Perin: 'Please buy the Jamadoba colliery. I am coming out to secure leases on other high grade coke bodies.'

The effortlessness with which the two worked together is a fine example of how a common purpose and vision can bring the most different of men together.

The buildings for the new plate mill alone totalled 1,15,000 tonnes of steel, fabricated in the US. The steel was bought in America at war prices. By 1917, the Tatas were looking for and encouraging industries that would consume the steel at the doorstep. For example, Burmah-Shell needed a lot of tin for its petrol cans and was willing to put down two-thirds of the capital for a tin plate factory in Jamshedpur. The Peninsula Locomotive Company came up as did the Hume Pipe Company, the Indian Wire Products, the Indian Enamel Company and some others. TISCO was barely ten years old when it was spawning its children and the town was developing at the same time.

Yet even the best laid plans can go awry when fate turns its back.

At the very period of its most ambitious extension it seemed as if all the stars were ranged against it. Prices spiralled in the wake of the war, there was serious shortage of transport in India and delays in delivery with labour trouble in America and Europe. The unkindest cut of all came from foreign steel companies, especially from Belgium where its iron works used the enormous scrap abandoned on the battle fields to make steel at cheap prices. Boat-loads of Belgium steel were dumped at Bombay and Calcutta and resultantly prices fell sharply. The profits of Tata Steel slumped from Rs 1.17 crore in 1921-22 to just Rs 1 lakh in 1922-23.

In July 1921, TISCO represented to the government that 'the situation was one of extreme gravity with about 40,000 employees and over Rs 10 crore of capital invested by some 30,000 investors, at stake'. It further pointed out that steel companies abroad had made huge profits during the war, and accumulated their resources, unlike Tata Steel which had voluntarily given 2,90,000 tonnes of steel at Rs 150 per tonne, much below the world prices or those prevailing in the open market in India. The government had benefited enormously. The company now advocated a policy of discriminating protection by the government since the steel industry was essential for national defence. Thanks to the support of Pandit Motilal Nehru and other leaders in the Legislative Assembly, a Tariff Board was appointed in February 1923 to look into the situation.

On the recommendation of the Tariff Board, the Steel Industry (Protection) Bill was introduced in the Central Legislation Assembly in May 1924. It was a historic occasion. For the very first time in history the British government was introducing a bill to protect an Indian industry. Sir Charles Innes, a member of Commerce, moved the Bill 'to provide for the fostering and development of the steel industry in British India'. Sir Innes said that one of the difficulties that confronted the government was 'that the steel industry in India is represented by a single firm'. Sir Chimanlal Setalvad presided over the debate and a motion was moved that it be referred to a select committee. The names proposed for the committee were those of Vithalbhai Patel (later Speaker); the eminent labour leader N.M. Joshi; Mr Chaman Lal; Pandit Madan Mohan Malaviya and Mr M.A. Jinnah. As soon as Jinnah's name was proposed, he rose to say that he had some financial stake in the company but if it was all right with the House, he would work on the committee.

Vithalbhai Patel asked: 'Is it a big stake?' Jinnah, an affluent lawyer, turned to Vithalbhai Patel and said: 'A big stake for Mr Patel, but not for me.'

Pandit Motilal Nehru, whose name was also proposed

for the committee, rose to say: 'I have no interest in the Tata Company. The only interest I had at one time was that Mr R.D. Tata had kindly put one of his motor cars at my disposal and I took good care of it...I had no other (interest) but do not despair of having some interest in future!'

As expected the bill was opposed by the representative of British trade interests in India, represented by the Associated Chambers of Commerce. It was also opposed from the other extremity of the House. Chaman Lal proposed nationalization of TISCO as did N.M. Joshi. Joshi insisted to put in a clause for the protection of the workers.

Jinnah fought for clarification. He said: 'I want this house to carefully grasp the issue and not be led away by extraneous considerations.' The bill, he said 'embodied a very important principle of state policy. For the first time the protection was offered to an Indian industry. It is a mere accident that the Tata Iron and Steel Company is directly and very naturally affected.'

After a heated debate, the bill was passed by a voice vote in June 1924. A bounty was given on the output of railway lines and fishplates for three years and an import duty of 15 per cent was imposed on imported steel. The company had been struggling against great odds for a long time and it seemed that time was finally running out. As if other challenges were not enough, an earthquake struck Japan, TISCO's principal customer for pig iron.

The company was reeling under these blows when a cable arrived from Jamshedpur saying that there was no money for wages.

Sir Dorab and R.D. Tata went to the Imperial Bank and Sir Dorab pledged his entire personal fortune of Rs 1 crore, including his wife's jewellery, to raise the Rs 2 crore that were needed at that time for a public limited company with 30,000 shareholders. Apart from the Greater Extension Programme, the cost of which was three times more than the estimate, there were debentures to be repaid. There was a period in 1924 when a good friend of R.D. Tata would call on him every day to ask when he was going to close down the works. Each day R.D. would reply: 'Ask me again tomorrow. We will be able to manage for today.'

It is difficult to imagine today that Tata Steel once faced such a grave crisis. What is particularly admirable is the resolve of these men who fought with courage and conviction for their dream risking financial crisis and perhaps even ruin.

At a meeting of the directors, someone suggested that the government be asked to take over the Tata Steel. R.D. Tata sprang to his feet, and, pounding on the table, declared that that day would never happen as long as he lived.

Fortunately by the end of 1924, production had risen by two-and-a-half times and the first returns from the increased production and the benefits of the Steel Protection Act were coming in. The Tatas were safe. However, it had been touch and go.

Throughout this baptism by fire, the Tatas honoured their obligations. There was rationalization but no retrenchment. Every penny owed was paid on time. R.D. Tata had to plead with the shareholders not to push for dividends. Speaking to shareholders in October 1923, R.D. Tata said:

'We are constantly accused by people of wasting money in the town of Jamshedpur. We are asked why it should be necessary to spend so much on housing, sanitation, roads, hospitals and on welfare...Gentlemen, people who ask these questions are sadly lacking in imagination. We are not putting up a row of workmen's huts in Jamshedpur—we are building a city'.

The shareholders, though restless, had remarkable confidence in the Tatas. And though, in thirteen years, they received their dividends only once, most of them held on to their shares.

Throughout the 1920s and even after the Greater Extension Programme was completed, a steady expansion continued through the 1930s. In 1939, when the Second World War broke out, Tata Steel was the 'largest integrated steel plant in the British Empire'—with a capacity of nearly a million tonnes.

6

National Leaders Intervene

As Tata Steel grew, it came to represent a new confidence that India and its people were beginning to feel. Tata Steel and the town of Jamshedpur had become a regular halt on the itinerary of many of India's key leaders and freedom fighters. While they fought for independence from the British, the Tatas laboured for a new industrial future for India. Many of these leaders were closely involved with TISCO and some of them even intervened when problems arose within the plant. The interventions of Subhas Chandra Bose and Gandhi were particularly remarkable. They helped resolve some of the key labour crises that arose during that time.

Gandhi's relationship with the Tatas goes back to 1909 when Ratan Jamsetji Tata (later knighted) sent him the first of five cheques of Rs 25,000. In 1910, with the second cheque of Rs 25,000 (equivalent of nearly Rs 14 lakh in 2006) a handwritten letter was sent which said:

'We, in India, must not forget that you and your fellow-workers in the Transvaal have suffered much and have sacrificed much to maintain our country's...[illegible] in the Transvaal, and that though your spirit might be steadfast your resources would be considerably diminished in so prolonged a struggle. Unless I feel therefore...[illegible] you receive renewed support it would be difficult for you to carry on so unequal a fight. Should you however be obliged to give up this struggle for want of due appreciation and support from us in India I feel it will be considered tantamount to an acknowledgement by us of our inferiority to the white races. What effect this would have in future in the treatment of our countrymen by the whites in various part of the world, could easily be imagined.

'Therefore I think it is the clear duty of all in India at this juncture to do what lies in their power—to give those who are engaged in this supremely important struggle the confident feeling that the vigorous sustained support, both material and moral, of their countrymen in India is behind them. If the cheque which I enclose herein will in any degree be instrumental in giving you and your fellow-workers this feeling, my object in sending it will have been accomplished.'

Gandhi wired Gopal Krishna Gokhale:

PRAY THANK MR TATA FOR MUNIFICENT TIMELY HELP STOP DISTRESS GREAT STOP PRISONERS' LOT HARD STOP RELIGIOUS SCRUPLES DISREGARDED STOP RATIONS SHORT STOP PRISONERS CARRY SLOP-PAILS FOR REFUSING STOP PUT ON SPARE DIET STOP SOLITARY CONFINEMENT STOP PROMINENT MOSLEMS, HINDUS, PARSIS IN JAIL STOP

Gandhi's association with the Tatas and Tata Steel went well beyond financial support. In 1925, he visited Tata Steel on the joint invitation of C.F. Andrews, who was then labour leader in Jamshedpur, and the directors. There was a dispute in Tata Steel between the union and the management and Gandhi was called in, in the hope that he may be able to arbitrate. Gandhi did.

Gandhi was received in Jamshedpur with much affection and excitement. He witnessed Charkha demonstration by women and in the evening addressed the mass meeting of over 20,000 people at the Maidan. Gandhi expressed his pleasure at visiting 'great steel works'. He said: 'As you know, I am a labourer myself. I pride myself on calling myself a scavenger, weaver, spinner, farmer and what not, and I do not feel ashamed that some of these things I know but indifferently. It is a pleasure to me to identify myself with the labouring classes, because without labour we can do nothing...My identification with labour does not conflict with my friendship with capital.'

He recalled the generosity of late Sir Ratan Tata in the struggle in South Africa and added, 'I wish to this great Indian firm prosperity that it deserves and to this great enterprise every success. And may I hope that the relations between this great house and labourers who work under their care will be of the friendliest character?'

His second visit was in 1934—this time, he was on his way to Orissa from Ranchi. Each time he was presented a sum of Rs 5,000 by the citizens of Jamshedpur. This time, it was for his Harijan work. Mira Ben accompanied him. During this visit, the president of the reception committee, Thakkar Bapa, mentioned a young girl was very keen to meet Gandhi, but having been taken ill suddenly had been moved to Tata Main Hospital. Even though he was in Jamshedpur for only ten hours, on hearing this, Gandhi took time out to visit the child in the hospital. After spending some time with her, he asked her permission to leave, to which she simply said: 'No'. Gandhi explained to her gently why he had to leave and after obtaining her permission he left the place—conferring his blessing on her.

However, that was not the last of the labour troubles

that TISCO faced. In the wake of the Russian Revolution, the 1920s were a time of labour turmoil and class struggle. TISCO, as the most prominent industrial plant, was not exempted. The steel experts imported from abroad were not refined in manners and tended to treat the workers roughly.

From the day of its opening in 1911 till 1920, things went smoothly in the Steel Plant although the covenanted foreign officers, in keeping with the times, treated the labour roughly. This was resented and disliked.

During the First World War , the workers were awarded a 10 per cent increase in wages. At the end of the war, they were nervous it may be withdrawn any day. While the management had no intention of withdrawing it, their communications with workers were very poor. They failed to sense this fear and reassure them.

The turning point in industrial relations within India came in 1920. The winds of the victory of the communist revolution in Russia blew over India, creating unrest. Literature in Indian languages, printed at foreign instance, flooded into the homes of industrial workers. The trade union movement was gathering momentum.

TISCO was way ahead of others in its benefits to workers, like an eight-hour day from the start and leave with pay, introduced in 1920, before the UK and the USA accepted it. However, because it was right in front as India's prime industry, it had become the target of attack. It was hit by a flash strike in 1920 that stunned the management. Another followed soon after in 1922.

A third strike took place a couple of years later, when the company was struggling against a threat of closure for lack of funds. The company had sacked the general secretary of the union, G. Sethi. Mahatma Gandhi, Dr Rajendra Prasad and C.F. Andrews came to Jamshedpur and discussed the recognition of the union and the case of Sethi with R.D. Tata. Sethi was reinstated and R.D. Tata agreed to recognize the union on the condition that it functioned along constitutional lines. The company went a step further and agreed to check off a 4-anna membership from the monthly salary on behalf of the Jamshedpur Labour Association. C.F. Andrews was chosen as the first president.

Mahatma Gandhi, with Dr Rajendra Prasad, on a visit to Jamshedpur.

C.F. Andrews tried to do the best he could for the union. However, his deep involevement in the national struggle for independence and other interests kept him outside Jamshedpur for long spells. And over a period of the next few years, a man called Maneck Homi, who was not even a member of the executive committee of the union, obtained the support of the workers. He spoke the language of the adivasis and was very popular with them. From February to May 1928, workers again downed tools at different times in different workshops. They had no demands. All they wanted the management to do was to talk to Maneck Homi. On 1 June, the company declared a lockout. At this stage, Subhas Chandra Bose assumed the presidentship of the Jamshedpur Labour Association from

C.F. Andrews and tried to mediate a settlement. Meanwhile, the management had also recognized the Jamshedpur Labour Federation of Maneck Homi.

The management signed an agreement with Subhas Chandra Bose in September which Maneck Homi denounced from the word 'go'. It was a very difficult time both for Bose and the management and during this period, Bose wrote quite a few letters to the management and also issued a stirring press statement assailing Maneck Homi, who, he thought, at first, was a friend of labour. He thundered to the press:

> 'By dividing labour, Mr Homi is serving neither labour nor capital. It will be an evil day for India if the whims and manoeuvres of one individual are allowed to undermine a public cause. Jamshedpur labour has to be saved and in order to save labour, the Indian Steel Industry there has to be saved from bankruptcy.'

The well-known labour leader, Mr Jamnadas Mehta, also intervened and came to the same conclusion about Maneck Homi as Subhas Chandra Bose had.

In a letter to Nowroji Saklatvala, who was to succeed Sir Dorab Tata as chairman, Subhas Chandra Bose wrote on 30 September 1928: 'Homi is still active. After the settlement, he raised the cry that we had been heavily bribed by the Company.' Subhas Chandra Bose quotes the instance of a worker of his who was first tempted, then threatened and finally assaulted by Homi's men. 'Unfortunately,' wrote Bose, 'we could not resort to such tactics because while Homi wants to gain his object by hook or by crook, we want to educate the masses. Consequently, for us, "honesty is the best policy".'

In a letter to L.A. Alexander, the general manager of the company, Subhas Chandra Bose states:

> 'My only interest is to save the labour organisation here from sinister and unscrupulous agents and to

give the Company an opportunity of readjusting itself and redressing the grievances of labour.'

Government officials discovered that Maneck Homi was guilty of defalcation of the union's funds and on that charge, Homi was imprisoned and prevented from trade union activities for a spell. He returned from jail to continue his activities but after 1945, he faded away from the scene.

Meanwhile, Subhas Chandra Bose who became president in 1928, stepped down in 1936 for health reasons. He did remarkable work during that time and pushed for including more Indians in the work force. His notable contribution was recognized by the Tatas; in a confidential letter written to him on 3 November 1928, Nowroji Saklatvala said:

'You may take it from me that nobody appreciates more than me the work you have done for the Tata Iron and Steel Company Ltd. and the sacrifice of your time that you are cheerfully making to see us through...It is only due to your personality and the sacrifice you are making in the cause of the Indian industry, that our Directors have sanctioned the further increment that you have suggested...India has no single concern of this importance and magnitude and it is your vision of the industrial importance of your country in its struggle for existence amongst other more powerful competitive nations, which, I am sure, makes you take this deep interest. Between us, we can and must make the industry prosperous.'

The succession to the office held by Subhas Chandra Bose in the Jamshedpur Labour Association was important enough to involve leaders of the Indian National Congress. The Congress leaders selected Professor Abdul Bari, a leading figure of Bihar and later, president of the Bihar Pradesh Congress Committee. Prof. Bari was an inspiring figure and he brought many young men into the trade union movement, including a youngster called V.G. Gopal, later to occupy the same office. But Bari was a firebrand and often intemperate in his language. He assailed the management in public in no uncertain terms and at one point, Jawaharlal Nehru and Dr Rajendra Prasad gave a report which took note of the intemperate words of the professor. In spite of their being at opposite ends, Bari seemed to get along with J.R.D. Tata. J.R.D. Tata told this author that the morning after his fiery speech slamming Tata, Abdul Bari would tell JRD that he said it for creating an impact, and he got carried away before the crowds. The manner in which JRD spoke of him to the author, he obviously understood Bari and was even fond of him. Meanwhile, before Bari came on the scene, V.V. Giri (later president of India) in 1932 started the Metal Workers' Union with encouragement from the management, but he had to give it up under pressure from Congress leaders.

In 1947, as Bari was rushing to meet Abdul Ghaffar Khan, the Frontier Gandhi, in Patna, Abdul Bari's car was stopped at the check post and armed police asked him to open the boot of his car. Prof. Bari lost his temper and engaged in a physical scuffle with the armed guard and, in that process, a rifle went off and killed him. Bari had given the union a sound base of support among the workers. He had publicly declared that his successor in office was Michael John and it was under Michael John, that the union and the company together pioneered something new for the entire country.

7

Union Stands Up for Management

Before Michael John became president, the chairman of the company, J.R.D. Tata, who was convalescing at a hill station after double pneumonia in 1943, did some hard thinking on TISCO's labour relations. He sent around a note in which he said:

> 'If our operations required the employment of, say, 30,000 machine tools, we would undoubtedly have special staff or department to look after them, to keep them in repair...But when employing 30,000 human beings, each with a mind and soul of his own, we seem to have assumed that they would look after themselves and that there was no need for a separate organisation to deal with the human problems involved.'

It is this note which provided the basis for the Personnel Department of TISCO that was established in 1947. Even before the department opened in 1947, and also after,

TISCO created many 'firsts' in India's labour history, which were often decades ahead of their time. A selection of these is given in the table. Most of these labour benefits were given by TISCO more than two decades before they were enshrined as law.

A list of worker benefits introduced by Tata Steel

Worker Benefits	TISCO Introduction	Enforced by law	Legal measure
Eight-hour working day	1912	1948	Factories Act
Free medical aid	1915	1948	Employee's State Insurance Act
Establishment of welfare department	1917	1948	Factories Act
Schooling facilities for children	1917		
Formation of Works Committee for handling complaints concerning service conditions and grievances	1919	1947	Industrial Disputes Act
Leave with pay	1920	1948	Factories Act
Workers' provident fund scheme	1920	1952	Employee's Provident Fund Act
Workmen's Accident Compensation scheme	1920	1924	Workmen's Compensation Act
Technical institute for training of apprentices, craftsmen and engineering graduates	1921	1961	Apprentices Act
Maternity benefit	1928	1946	Bihar Maternity Benefit Act and by Govt of India
Profit sharing bonus	1934	1965	Bonus Act
Retiring gratuity	1937	1972	Payment of Gratuity Act
Ex-gratia payment— road accident while coming to or returning from duty	1979		

Two agreements were signed in 1956, especially a supplementary one in August, which proved to be the turning point in industrial relations. These are significant as they laid the foundation of modern industrial relations in Indian industry. This agreement guaranteed the rights as well as obligations of the workers and both were spelt out, the workers being guaranteed a security of tenure.

The signing of the agreement in 1956 which paved the way for modern industrial relations in India.

At JRD's instance, a process of joint consultation between management and the union was initiated, and the union collaborated with the management in those areas where the management's decision had a bearing on the welfare of the workers. Forty-two Joint Dialogue Councils were developed and sixty-eight Joint Consultative Committees dealing with various fields of activities from canteens to safety and health were formed. The Joint Dialogue Councils did not touch on the basic wage structure, which was negotiated on an all-India level, as this agreement involved government steel plants as well. Short of that, it involved everything including bonus. This was perhaps the finest innovation in labour relations in the history of industrial India and continues to be active even today.

Supposing, today there is a case against a worker, how is it dealt with? Over the years, procedures and rules have been worked out and agreements arrived at between the union and the management. If it is a case of theft and it is proved to the satisfaction of the union, the worker is dismissed and the union will not thereafter take up the worker's cause. In the event of any misconduct on the part of the worker, the management holds an inquiry and takes all the normal procedures under the Labour Act. The worker can appeal against the management's decision to one of the seven Zonal Works Committees, where the management and the union are represented—the management chairing the committee with the union representative. The worker has a right to a second appeal, if there is a difference of opinion within the Zonal Works Committee. Then, the Central Works Committee hears the second appeal where five representatives of the management and the union consider the case. In case there is a difference of opinion there, the union president takes it up directly with the vice-chairman of the company. In the last seventy-five years, in the largest private work unit in India, not a single case has gone up to a labour court or to government arbitration.

J.R.D. Tata was not the only one to be proud. As the president of the union, V.G. Gopal, said, 'The 100 per cent efficiency is on the rated capacity that is known to everybody and not one that is artificially scaled down to what is manageable. Whenever I meet Mr Tata, he always enquires: "into what new areas are you expanding the consultative process?"' When asked, what is the difference between dealing with TISCO and dealing with other companies, Mr Gopal said, 'a lot of difference. Once I convince TISCO, the issue is immediately settled. In other companies, it does not happen so and implementation takes longer. If we have a difference of opinion and the Union's contention is right, the Management accepts. If the Union's contention is not right, the Union will not press the case.'

At the Platinum Jubilee celebrations, the then prime minister, Indira Gandhi, said in Jamshedpur that Tata Steel was a 'beautiful example' for others to follow for sustained industrial growth of the nation.

When this writer asked J.R.D. Tata why, with less than 3 per cent government investment in Tata Steel, the Tatas still opposed any government attempt of nationalization in 1978, he replied that had the government taken it over, the unique labour relations that have been built up within the company would have ended. As a result, something very precious would have been lost.

In 1982, 81 per cent of the holding of the parent company, Tata Sons, went to charity through various Tata Trusts[1] (less than 2 per cent to the Tata family and the directors). 'If the employer is a trustee on behalf of society,' said Gopal, 'so is the worker. Only when both are ready to accept the broad rules can there be room for negotiations.'[2]

In October 1979, the company celebrated fifty years of industrial harmony between the management and labour; in those fifty years, there were only two spells of stoppages, neither of which were aimed at the management.

The first was a political stoppage in 1942 when Gandhi gave a call for 'Quit India'. The second was in 1956 when the communist party led by S.A. Dange made a major bid to take over the leadership of the workers front, the Congress-affiliated Tata Workers' Union. 'After three weeks, the communist bid fizzled out because,' says Gopal, 'the Company stood firm and the workers realized the issues at stake.'

In 1978, two cabinet ministers in the Janata government made a bid to nationalize Tata Steel. When the first

---

[1]Tata Trusts hold 66 per cent share in Tata Sons in 2007.

[2]From Manohar Malgonkar's *Cue from the Inner Voice*, Vikas, 1979.

rumours were heard, the union sent a telegram to the prime minister, Morarji Desai, on 4 October 1978, which read:

> 'Tata Workers' Union Executive Committee seriously disturbed over press reports that Central Government contemplating nationalisation of TISCO. TISCO's performance has been the best of all steel plants. Labour relations also most cordial over last four decades. In the circumstances, workers most unhappy over Government's contemplated move. Request utmost consideration of our above points before taking firm decision.'

At the Janata National Executive meeting on 28–29 December 1978, then industries minister, George Fernandes, and steel minister, Biju Patnaik, formally submitted a proposal for the nationalization of key sector industries which were profitable in steel, aluminium (Hindalco) and automobile production (Telco). Particular attention was focused on Tata Steel. The gauntlet thrown by the steel and industries ministers was first picked up by the executive of Tata Workers' Union. They met on 7 January 1979, and noted with deep concern the fresh wave of statements by union ministers. The workers' union in a resolution observed that they were pioneers in the field of collective bargaining, joint consultation and closer association of workers in the running of this company and noted that 'the views of the overwhelming majority of the employees of the company through various Joint Departmental Councils (were) against nationalisation'.

The union resolved:

> '[...] THAT the Prime Minister, the Steel Minister and their other Cabinet colleagues be requested to take note of the workers' reaction against the nationalisation of the Tata Iron and Steel Company Limited, because this Company has been maintaining a very high level of production, cordial industrial

relations over the past several decades which is an envy for the rest of the country and where the participation in management is a way of working.'

The resolution added that nationalization would lead to a loss of production and 'it will be detrimental to the interest of the nation'. The next day, the chairman of the company, J.R.D. Tata wrote to all the members of the Janata National Executive. Many cabinet ministers as well as the press and public opinion were overwhelmingly against nationalization.[3] But it was the union that took the wind out of the sails of the nationalization lobby.

[3]*INDIA says 'No' to Nationalisation*, R.M. Lala, The Rajaji Foundation, 1979.

Part III

The Spirit of Adventure

8

Men of Steel

Jamsetji Tata was the visionary who laid down the guidelines for TISCO. He not only found the finest consulting engineers in the world, but also instructed his son Dorab that the works had to start with Americans at the blast furnace, the Welsh at the coke ovens, Germans at the open hearth furnace and the British at the steel rolling mills. When the works opened eight years after his death, the primary composition of nationalities was as he had instructed.

The founder of the House of Tatas had envisioned an industrialized India of which the major constituent was steel. It was this vision which guided him and those who followed to search endlessly and endure many disappointments before they tasted success.

Interesting enough at the other end, in eastern India, P.N. Bose whose explorations steadily pointed the way for location of the steel plant, had the same objective. For Bose was far more than a geologist. He organized the Second Industrial Conference in Calcutta in 1906. He also founded the Bengal Technical Institute, in 1906, and was its Honorary

Rector for the first two decades. In his later years, he rejoiced to see TISCO become the largest single employer of his students.

Unknown to many, Bose was also a historian and wrote books. His interest in history and society is apparent when one considers two of his well-known works: *Epochs of Civilisation* and *Hindu Civilisation under British Rule*. However, the most remarkable thing about Bose was his humility. Whenever complimented on the discovery of Gurumahisani, he humbly replied: 'Well, to compare small things with great, I discovered them as Amerigo Vespucci is said to have discovered America. All he, and Columbus a few years before him, did was to bring it to the notice of the Europeans. The iron ores of Mayurbhanj had long been worked by smelters of the State. All I did was to make them known to the industrial public.' The Tatas, in appreciation of his extensive contribution and pioneering work, erected his statue near the directors' bungalow, Jamshedpur and endowed a Chair in his honour at Tata College, Chaibasa, now in Jharkhand.

Charles Page Perin's association with Tatas extended for almost thirty years from the day Jamsetji walked into his New York office. Perin was repeatedly called to India. And even after the Greater Extension Programme was completed in 1924 he remained the primary expert whom the Tatas consulted. Bose recalled that when Perin and C.M. Weld went to explore the interior of Mayurbhanj with him:

'We had to do hard work in the jungles either on foot or on elephant and as they were State Guests, I provided beer and other alcoholic drinks for them. But all the time we were engaged in the jungles, they would not only abstain from such drinks but eschewed even such an exhilarating mild stimulant as tea. They would drink only soda water'.

Bose called Perin 'one of the most level headed businessmen I have ever come across'. Perin came from the West, Bose from the East; the twain met in Mayurbhanj and learnt to respect each other. When Kipling wrote 'East is East and West is West and never the twain shall meet', he also added:

> But there is neither East nor West,
> Border, nor Breed, nor Birth,
> When two strong men stand face to face,
> Though they come from the ends of the earth

These lines quite appropriately describe the relationship that existed between Perin and Bose. They were both men of strength, purpose and courage. Their goal rose above all differences of nationality or culture to take the form of a camaraderie built on mutual respect and admiration.

Jamsetji's genius was to inspire his sons and his colleagues to continue his work in the years after his death. Dorab had earlier trundled in bullock carts with Weld, and often slept in them at night because it was unsafe to pitch even a tent in the jungles. Later he rode majestically on elephants through the forests of Mayurbhanj.

When Mahatma Gandhi said that the Tatas represent the spirit of adventure, perhaps he had Sir Dorab in mind too. Dorab, knighted in 1910, used to say: 'Kind fate has, however, prompted me to help in bringing to completion his (Jamsetji's) inestimable legacy of service to the country.' Jamsetji's younger son Ratan is featured in the early photographs of Sakchi. He was a clubbable man who was knighted in 1916. When travelling by ship to England, his liner was torpedoed in the Mediterranean. He died in England in 1918. At the crucial crisis of TISCO in 1921–24, it was Sir Dorab and R.D. Tata who wrestled each day for its survival.

In 1904, R.D. Tata was at Jamsetji's bedside in his last days and reminded him of the honour he had brought to the family's name. 'If you cannot make it greater,' Jamsetji said, 'at least preserve it. Do not let things slide. Go on doing my work and increasing it, but if you cannot, do not lose what we have already done.' This was Jamsetji's last message.

The way in which R.D. Tata conducted Tata Steel day after day, not knowing what the morrow would hold, showed grit and loyalty. He was the main contact with political leaders like Motilal Nehru in the Legislative Assembly, who, in concert with the British government, raised tariff against imported steel and gave other financial incentives to TISCO. However, the cumulative strain of five years must have told on his health and he died in 1926 having steered the company to safer waters. Seth Jamnalal Bajaj, the businessman who dedicated his entire wealth to Mahatma Gandhi's wishes, wrote of R.D. Tata: 'If all businessmen in India would acquire half his love for things Indian, there is no reason why all our enterprises should not flourish.' The *Indian Mail* in an obituary observed: 'His courage was such that those who worked with him were transported into heights of the greatest altitude even when they knew they had reasons to be despondent.' The paper called him a man of 'regular and steady habits and moderate living, upright and following the highest biddings of his own conscience, courageous like a lion when occasion demanded, yet gentle in ordinary life, with a high sense of duty'. 'His career,' the *Indian Mail* concluded, 'has been a beacon light to all those who came in contact with him.'

Burjorji Padshah, along with Sir Dorab and R.D. Tata, formed the undaunted trio that led Tata Steel through its trials and tribulations at various points in its early history. Padshah's finest hour was at the start of the company and the early years when he was executive director. His faith and confidence in Jamsetji's projects was contagious.

Padshah (as noted in Chapter 6) was a bit of an eccentric but not the only one. A Britisher who had lived in central Africa and worked for years in Jamshedpur insisted on living in his pitched tent. When he went to bed at night it was unlikely anybody would approach him for curling up at the foot of his bed, was his pet—a fourteen-foot rock python.

In Gary, Indiana, USA, is the plant of the US steel company that supplied—willingly or unwillingly—a good number of top technicians to TISCO. Among the first was Temple W. Tutwiler. 'Terror', rather than 'Temple', would have been a more appropriate name. A huge, ferocious man, he could swear like a trooper—and did. As the first general manager of the works he made sure everyone obeyed and recognized him. Nobody dared overtake his car or turn up at the works in a tie.

Temple W. Tutwiler, the rough-mannered but highly efficient and ingenious general manager of the plant.

When Lord Chelmsford put out his hand and said: 'I presume you are the General Manager,' Temple W. Tutwiler shot back: 'You're Goddam right'. It was taken as an affront by the representative of His Majesty the King and Sir Dorab's secretary had to write to Tutwiler suggesting that he apologize to the viceroy. Tutwiler went up to Delhi, explained to the viceroy his rough background and put things right. And so, Anglo-American relations became normal again!

Every Christmas, the directors met in Jamshedpur for their board meeting. They played poker with Tutwiler and even beat him at the American card game. Tutwiler may have lost in a game of poker but never at his work. When he was first asked to make steel shells in 1914 the plant was two years old, and what he had were open hearth furnaces. Shells had never been made before on open hearth furnaces but he took a deep breath and went to work. The Tatas had no way to press them so Tutwiler ransacked every railway and ship-building shop that had a lathe to bore five-inch rounds into shells. It worked and saved the day in Mesopotamia for British and Indian troops. He was succeeded in 1926 as general manager by C.A. Alexander, another American.

John Kennan, who became general manager in 1931, was a contrast to Tutwiler though he, too, was from Gary. Keenan was courteous and affectionate to all the workers. It was he and Sir Ardeshir Dalal who completed the Indianization of the works in the 1930s.

Keenan was a gifted raconteur who gives a vivid description of life in Jamshedpur, 1913–37. He relates in his book, *A Steel Man in India*, how, when he first sailed in 1913, 'all the old-timers on board were mightily amused that I was going out to India to make steel in an Indian concern. Indians could weave cotton cloth and make jute gunny bags but steel! Why even the Tatas were sure to close down within a year.' When he came to Sakchi there was one blast furnace making 200 tonnes a day. Some years later, Keenan proudly noted that in the whole of the USA only two blast furnaces improved on the Tatas' 1,100-tonnes-a-day furnace. Soon after he came he found the Indians so adept at their jobs that twenty out of the twenty-eight Americans were shipped home ahead of time, within eighteen months of their arrival. 'I saw,' he notes, 'the transformation of the old cultural India into the new industrial India, with Tatas as the backbone...' Keenan stayed twenty-five years in Jamshedpur and developed a wide circle of friends. One of his guests in Jamshedpur was the American author Louis Bromfield whose novel *The Rains Came* was set in India. Bromfield called Keenan 'the Irish-American Maharaja' who 'liked people, and whisky and horseracing, and most important of all, all kinds of Indians...He could talk a mile a minute and was a great story-teller...Never before or since have I encountered such vitality.'

The success of the plant naturally, also, had a positive effect on the ancillary businesses that supported the main steel works. It spurred many a success story. In 1910, a Sikh called Indra Singh came into town and asked for a job. His ambition was to be a construction contractor. He learnt that the only vacancy was at Gurumahisani. So he walked the 45 miles to take up the job at Re 1 a day, with the right of becoming mining contractor for the company later. Indra Singh picked his labour carefully and paid them slightly more than other contractors. His teams for picking ore were by far the best pickers and soon the company learnt that they could depend on him in any crisis. Some years later he started the Indian Steel and Wire Products and before too long he became Sir Indra Singh. His firm was run by his son Ajaib Singh till 1990 and closed down for ten years. Tata Steel took it over and with fresh inputs of expertise and equipment it is more productive than ever before.

John Peterson, a former controller of munitions and a retired I.C.S. official joined, TISCO in 1921. He handled the company's affairs for ten years. As managing director, based in Bombay, he was quick at decision making, and responsive to any good proposal put to him. When the company was going through the financial crisis of the 1920s, Peterson, in the Bombay head office, Perin, in New York, and Alexander, as general manager in Jamshedpur, made a crack trio. A new spirit of achievement permeated the company and all department heads worked as a team to bring production to its peak.

In 1925, when R.D. Tata brought his twenty-two-year-old son, Jehangir and requested Peterson to train him, Peterson promptly ordered an extra desk in his room and settled the young man down. Then on, the young JRD was at every interview and saw every paper going to Peterson's desk and returning from the managing director's desk. JRD recalls Peterson as a Scotsman especially kind to Indians and exceptionally rough with his own kind. Like Keenan, Peterson too reacted negatively to the superior attitude of many Britishers of the time in India. It was Peterson who, apart from shaping JRD in administration, passed on to the young man his love for poetry.

After a few months under him, Peterson sent JRD to Jamshedpur for practical exposure. JRD's bible was a book called *The Shaping of Steel*. In 1931, Peterson was succeeded as managing director by a cigar-smoking strong man, Sir Ardeshir Dalal, I.C.S. (retired) and a J.N. Tata scholar in his younger days. Sir Ardeshir made a substantial contribution to placing the company on a sound footing and TISCO turned the corner under his care. As he hastened the process of Indianization he made sure that the efficiency level did not suffer. For a brief spell, during the Second World War he was on the viceroy's executive council in charge of planning. It was he who planned for a string of India's national laboratories.

Jehangir Ghandy

Jehangir Ghandy, a double graduate in chemistry, was interviewed by Tutwiler in 1919. When asked, Ghandy replied that he expected to work in a laboratory. Tutwiler grunted that he wanted none of this book learning and asked him to report at the coke ovens at 6 a.m. next day. Next year, Ghandy joined the blast furnace as extra foreman. He was sent to Columbia to qualify in business administration and to the Carnegie Institute of Technology to study metallurgy.

In 1930, when Jehangir Ghandy became the first Indian general superintendent, a sadhu walked 400 miles to meet him and requested him for a bell for his temple. Sir Jehangir ordered the works to build the biggest temple bell in India—of 760 lbs—and forgot all about it. Eight years later, when Jehangir Ghandy became the first Indian general manager, the sadhu turned up again. This time, only to congratulate him.

Former finance minister and Tata director, Dr John Matthai, called Sir Jehangir 'the ablest administrator that has been thrown up in recent times'. The government drew liberally on his talents for national planning and technical education. Sir Jehangir was a warm-hearted man with a twinkle in his eyes. When the Second World War broke out, he became the man responsible for the War Fund in Jamshedpur. He raised money for a Spitfire for the Indian Air Force and another Spitfire to help the Royal Air Force in the Battle of Britain. Lord Beaverbrook, munitions minister, wrote to Sir Jehangir expressing: 'Heartfelt thanks for a gift that brings to all of us in Britain encouragement and inspiration for the stern tasks that lie ahead.'

One of the lesser known heroes of the Second World War was a man called Dhanbura. He was sent by the Tatas to the Broken Hill Proprietory Company, Australia, to learn advance technology of armoured plating for cars. Dhanbura worked out his own improvement on the technique of armoured plating and sent back the advanced technique to assist Australia. In doing so Dhanbura was among the first to give TISCO's expertise outside.

Another young man who followed in Dhanbura's steps thirty years after the War was M.N. Dastur who had joined TISCO as a worker. He later became a steel consultant setting up or advising on steel plants, in Libya, the Gulf and Latin America. It was the dream of Sir Nowroji Saklatvala, who succeeded Sir Dorab as chairman in 1932 that a day would come when Indians would give their expertise abroad as Germans were doing in the late 1920s.

On Sir Nowroji's untimely death in France in 1938, the directors of Tata Sons, all senior seasoned men, chose from among themselves their youngest colleague, J.R.D. Tata, to be the chairman of Tata Sons. With this appointment, he was also elected the chairman of Tata Steel.

Tata Steel had on its board over the years Sir Cowasji Jehangir, Sir Fazalbhoy Currimbhoy, Sir Purshottamdas Thakurdas, Sir Chunilal B. Mehta, F.E. Dinshaw, Sir M. Visvesvaraya, Shapurji Bharucha and members of the Thackersey, Goculdas, Rahimtoola, Mafatlal and the Khatau families, among others. Sir Chunilal was director for forty-nine years from 1911 to 1968, with a break of eight years when he was in government. Sir Visvesvaraya, the industrial wizard from Mysore, served for thirty years on the board and at one time was one of the few technical men on the board. Keshab Mahendra served on the board for thirty-seven years. S.A. Sabavala, as vice-chairman, has played a vital role for decades.

Russi Mody

As head of the mining section, Russi Mody ensured facilities which propelled Tata mines to the very top in production and morale.

Educated at Harvard and Oxford, Mody joined TISCO in 1939. As managing director from 1974, he sat in different workshops each week, meeting workers, accepting petitions and signing personally every reply to the grievance or request of a worker. 'In ninety percent of the cases there is nothing I can do, but the fact that they hear from me means something to them,' he said. Well-formulated rules and guidelines were in place but TISCO also gave scope for a worker's appeal to be heard at all levels. If anything could be done, the worker was confident it would be.

TISCO elicits a loyalty and carries on a tradition handed down from father to son, extending over four generations and now the fifth generation is just coming in. A TISCO employee after twenty-five years service could suggest his son, son-in-law or a relative for a post. This person was put on the register and at first would get a temporary job. And at the next suitable vacancy he has priority. If the employee completed forty years, the company guaranteed his relative a job. This system may not always make for efficiency (although the company gave training) but it did create a deep sense of loyalty and trust that cannot be measured by material standards. This tradition of employing people from one generation after another from the same family continued till quite recently. However, things had to change after India entered the age of globalization and found itself floundering in a world of competitive markets. It fell to Russi Mody's successor, Dr Jamshed Irani, to undertake this delicate task.

Steel men live hard, work hard and risk their lives each day. Keenan records that one day, in the 1920s, a mixer crane was hoisting a ladle containing 75 tonnes of molten pig iron:

'Nobody knows just what went wrong. But slowly
at first the ladle fell away from its supports, then

severed itself completely and crashed to the ground, vomiting sparks and burning metal. As the molten iron reached out toward the brick-layers it met and exploded puddles of water. The air was filled with the rending noise of the ladle flinging itself to earth, the confused and frenzied shouts of the men, and the inimical hiss of steam.'

Eleven men were writhing on the ground with burns, their beseeching eyes crying out for help. The ambulance could take only five and Keenan had the difficult job of selecting the next three for his small car.

'I spotted one who had more skin than the others, who could turn his head and follow my movements. Turning to the rescuers, I said in Hindi, "Take him".'

'The man shook his head in negation. "Do not take me away," he said. Turning his head feebly, the Hindu nodded toward the body of a half-burned Mohammedan whose chest was heaving in agony and spoke. "*Hamara bhai ko lejao*"—Take my brother, he said clearly. The Hindu who was in pain and in danger of death remembered, not that the Mohammedan was of a different faith, but that he was his brother.'

This unknown Hindu, who for lack of immediate succour perhaps died, was the most towering of the Tatas' 'men of steel'.

9

Growing Younger

The two angular arms of Howrah Bridge at Calcutta are thrown over the spacious Hooghly river. Its magnificent silver lines reflect the rays of the sun. Upon this bridge, for over seventy years, millions have travelled each day by foot, rickshaw, two wheelers, cars, trams or buses. Fabricated almost entirely from 'TISCROM Steel', it stands as a lasting monument of beauty and strength; and testament to the skill and ingenuity of Tata Steel. The specifications for the project in 1930s called for high tensile structural steel capable of bearing a load of 43 tonnes per square inch. There were many difficulties in developing this steel and even after the Tatas had developed it, they had to contest a patent suit to prevent the use of imported high tensile steel, before bridge engineers would accept TISCROM.

In the 1930s, Tatas also developed TISCOR, specially designed for safe welding in structural steel. It is used in ship construction, trolley buses and freight cars. With the outbreak of the Second World War in 1939, great demands were made on the ingenuity of TISCO scientists and in the course of the next five years, they managed to produce 110 varieties of steel with hardly any worthwhile facilities available anywhere in the country except in their plant. The Tatas' major achievement was armour-plating for cars. After experimenting with different compositions of steel, and after testing them on firing ranges, the Tatanagar, a light armoured vehicle, was put on the war front of north Africa. A press item in the 1940s was published abroad under the headline 'India-made Armoured Cars Praised'. It said:

> '"Safer than slit trenches during a bombing raid", was a gunnery officer's tribute to the cars doing service in the 8th Army. An officer goes on to describe how a 75 mm shell burst on one side of Tatanagar. The metal plates were buckled but nowhere pierced. The four occupants of the car emerged unscathed...Units possessing [the] Tatanagar swear by them.'

The Master-General of Ordnance, Simla, certified the Tatanagars as 'excellent'.

The Tatanagar, a light armoured vehicle manufactured by TISCO that was used during the Second World War.

At a crucial time during the war, nickel, an essential ingredient in armour-plating, was in short supply. The Tata metallurgists created a new composition of armour with the same good ballistic properties but with a third of the nickel content, thus working around the problem of the short supply of nickel. They also succeeded in making magnetic steel bars, and special or high-speed tool steels. In creativity and innovativeness, the Second World War was TISCO's finest hour. After Independence the same skill was used to produce in TISCO's axle plant, turret rings of the Vijayanta tanks on which its gun carriage sits and revolves.

Incidentally, in view of Jamshedpur's supreme importance in the war efforts, the British decided to smoke-screen the whole town to camoflage it from Japanese bombers in Burma. Several crude mortars and boilers were lighted to belch black soot. The staff became 'sooted' and no planes came near Jamshedpur.

The ingenuity of Tata scientists extended to an itch to make old machinery, or even scrap, productive. A few decades ago, Tata Steel purchased four scrap locomotives from the Bokaro Steel Plant. These were not even in a condition to make their own way to Tatanagar railway yard and had to be freighted on railway wagons. TISCO's loco team set to work to revamp these locos and, before long, they joined the TISCO ranks of sixty locomotives that once chugged along over 200 miles of rails within the plant.

Till the early 1980s the TISCO plant had a precious assortment of locomotives at work and railway lovers in England even wanted to bring a party of thirty-five to visit them all the way to Jamshedpur. The then general manager, Dr Irani, replied: 'Hurry up. They won't be there for long.' The pride of place is occupied by the red painted Cheetah that chugs no more but is perched majestically on its rails.

Adversity brings out the best in men. During the 1960s to the 1980s, a time of severe government regulation, commonly called the Licence Raj, the ingenuity of the scientists at TISCO was fully stretched. An old generator was running at 85 per cent capacity. TISCO engineers tinkered with the machine for weeks on end and finally brought it to 100 per cent capacity. They did it because it presented a challenge to them. It was an apt intervention for an acute power shortage that struck a short time later.

Such skill is not born overnight. In 1911, when pig iron was first made, all the 200 technicians were foreigners. A decade later, the company established the Jamshedpur Technical Institute, to train future technicians. A chemical laboratory had been established even earlier.

The company did not rush headlong into Indianization without making sure that it had built itself a capable technical cadre at different levels of steelmaking.

By the mid-1950s, the Jamshedpur Technical Institute had trained 13,000 technicians and when the public steel plants came up this pool of technicians provided the talent required for manning them. Two thousand technicians were specially trained at the institute for the upcoming public steel plants as well as the heavy engineering units of the government. While innovativeness may affect savings, there are times when it may necessitate a heavy financial cost to the company. In 1979-80, there was a severe shortage of railway wagons of which TISCO required 1,000 daily. TISCO immediately secured a fleet of trucks to carry coal and iron ore at three times the cost to make sure that its ever-hungry blast furnaces were fed. When other steel plants suffered due to poor delivery of raw material, TISCO kept up its 100 per cent efficiency.

The initial capacity of TISCO was 72,000 tonnes of ingot steel. With the Greater Extensions, it rose by mid-1920s to 4,29,000 tonnes and by 1934-35 to 6,06,000 tonnes. TISCO fought its way through the depression of the 1930s with a programme of expansion. By the time the Second World War broke out in 1939, production was almost a million tonnes with the Tatas ranking as the largest integrated steel plant in the British Empire. At the end of the war in 1945, its machinery overused, the plant needed renovation and replenishment. Only after receiving assurances that there would be no steel nationalization, did the company plan a Modernisation and Expansion

Programme (MEP) and launched it in 1954, to take the capacity of the plant to 1.3 million tonnes of raw steel.

Not long afterwards, permission was obtained to raise the plant capacity to two million tonnes (of raw steel) and TISCO launched its Two Million Tonnes Programme (TMP). The MEP and the TMP got dovetailed into one programme. Again expansion meant not only putting up different mills for the plant but also the mechanization of the iron ore mines and the extension of the rail facilities in addition to the usual extension of town facilities.

The consultants for the TMP were Kaiser Engineers of Oakland, California. The World Bank gave the company the largest loan granted by it until 1957 to an industrial undertaking anywhere in the world. It was also the largest single loan granted for any purpose to any country in Asia. The expansion was completed in thirty months.

The policy of industrialization of the government spelt out in 1956 under the then prime minister, Jawaharlal Nehru, laid down that 'the commanding heights of the economy' would be the preserve of state enterprise. Tata Steel received no permission to expand till the mid-1970s based on which the first phase of modernization was completed in 1982.

The government had relented on two earlier occasions but unfortunately both times unexpected changes stalled the exapansion. In the early 1960s, the company received the permission and appointed W.S. Atkins of the UK as consultant for a 4 million tonne project involving the installation of a strip mill. While the company was in the midst of preparing a project, a major devaluation of the rupee took place in 1966 and sent the cost estimates soaring. The company shelved the project. Similarly in the early 1970s, Nippon Steel of Japan prepared a feasibility report for the expansion of the capacity of Tata Steel to 5 million tonnes. This time the fuel prices shot up in the wake of the Arab–Israel War of 1973, forcing the company to call off its plans.

During these thirty years, when the demand for steel was reasonably high, the government kept a retention price for Indian-made steel that was substantially lower than the prices of imported steel. This barely provided the company a 3 or 4 per cent return on its gross block. For thirty years, this policy was retained which made it possible for speculators in steel to make enormous profits while starving Tata Steel of legitimate profits it could have ploughed into expansion. Wisdom dawned in early 1991 when the government lifted the retention price order, giving the entire steel industry—both private and public— the spurt it needed and a fair price for its produce.

The main public sector steel plants started in the mid-1950s were at Durgapur, Rourkela and Bhilai. They were set up with British, German and Soviet assistance, respectively. They had the advantage of the latest equipment at that time. In spite of the TMP being completed about the same time as these public sector plants, a good part of the TISCO plant continued to be run on equipment dating to the 1920s and the 1930s. Even with this handicap, its performance was unbelievable, as the following comparative figures of productivity given to the parliament by the steel minister in 1979 indicate:

1) Indian Iron and Steel 63.3%
2) Durgapur 69.7%
3) Rourkela 96.7%
4) Bhilai 98.2%
5) TISCO 106.7%

If TISCO excelled at efficiency, it is because of the innovativeness of its staff, their morale and the zealous endeavour of the management to keep the plant young. It was, by no means, an easy task. American manufacturers who had originally produced the machinery had discontinued making spare parts for equipment they now ranked as antique. TISCO, however, had to carry on with its machinery because of the constraints on both price and expansion.

Fortunately, Tata Incorporated, New York, constantly purchased the steel machinery required to keep the TISCO plant in shape over the decades. Only with the Modernization Programme of the 1980s, did Tata Steel get a breather and the latest designs and technology of steel-making enabled it to say goodbye to a lot of the old machinery, and more important to obsolete technology.

One man whose fore-sight helped Tata Steel meet some of these challenges was Sumant Moolgaokar, the creative genius of Telco (now Tata Motors), who was also vice-chairman of Tata Steel from 1966 to 1979. In 1966, he decided to set up what he called a 'Growth Shop', not so much to serve the day-to-day needs of the steel plant which was being done internally, but to manufacture the capital machinery which would make the steel at a time when the availability of foreign exchange was a great impediment.

Sumant Moolgaokar

'A company,' said Moolgaokar, 'has to constantly grow younger. It has to grow just not in size but in productivity and technology. You have got to give the technicians the fun and excitement of growth. In a developing country, if you are importing a crane, it is no challenge; but if you make one, you give your technicians fresh skills. It is part of a manager's duty to give that challenge to his people.' Moolgaokar spoke from experience. As a young engineer during the war years, no imports of machinery were possible. He had then, produced an entire cement plant from scratch for the Associated Cement Companies (ACC). This brought him to the notice of J.R.D. Tata. When Telco was in a crisis in its early locomotive manufacturing period, JRD requested

Sir Homi Mody, chairman of the ACC, to part with Moolgaokar, which they reluctantly did. Moolgaokar, recounting his early years, told this author that he joined as general manager, 'Telco produced more general managers than locomotives!'

To create such a Growth Workshop, a first-class team of design engineers is even more important than machinery. Over the years, such a team was shaped at Adityapur, just outside Jamshedpur, where a township has now developed. The company has received offers from foreign firms to collaborate with TISCO in the supply of steel-making machinery from the 'Maintenance Shop' at Adityapur to other developing countries. But the prime job of the Adityapur Shop was to fabricate the enormous amount of machinery needed for its modernization, thereby saving the company—and the nation—a sizeable amount of foreign exchange.

Also at Adityapur, an entire colony of ancillary industries have come up primarily, though not entirely, geared to supply TISCO. Tata Steel farms out jobs to the ancillary suppliers, 80 per cent of them located in Bihar and Jharkhand. Tata Steel loans its ancillary suppliers the steel. If the quality of the components is sub-standard, the supplier is not penalized and made to pay for the steel. The sub-standard component goes into the company's scrap yard. The company has encouraged retired people and even some family members of their staff who cannot get jobs to start the ancillary industries. That is how so many ancillary industries have grown.

Maintenance involves training of men as much as the upkeep of machinery. TISCO was one of the first companies to launch a Staff Training Institute (later named the Management Development Centre), in 1948. Each year, it holds over a hundred different programmes training thousands of employees in a variety of areas.

The key factor in steel-making is the use of fuel. The Indian steel industry once accounted for 19 per cent of the petroleum consumption in the country and 14 per cent of coal consumption. Between 30 to 37 per cent of the cost at the steel works go into fuel. Just as an ageing car consumes more fuel than a new one, an ageing plant consumes more energy. Even under the circumstances, the Fuel and Energy Department of TISCO, started in 1930, managed to reduce fuel consumption in the 1970s by 10 per cent, at the same time, modestly pushing up its level of saleable steel. The department encouraged technical innovations but also launched a programme of Human Management with the slogan: 'whatever burns, never returns'. The steel melting shops which were operating with age-old open hearth furnaces, effected a saving of 18 per cent in energy consumption. With subsequent modernization, further fuel reduction came about.

The Modernization Programme, long overdue, was launched in 1980 with a loan of US $38 million from the International Finance Corporation. The Modernization Programme was undertaken in two phases. Phase I would raise the level of saleable steel from 1.52 million tonnes in 1981 to 1.74 million tonnes in 1983.

TISCO's raw steel production in 1981 was 2 million tonnes. Fortunately, as revealed later in this book, wiser counsel prevailed to modernize the entire steel plant in danger of becoming obsolete.

10

Precious Minerals, More Precious Men

Over centuries the adivasis called it Noamundi, 'That Mountain'. They knew that there was something special about it, but they could not have been aware that on that mountain rested 200 million tonnes of iron ore—one-thirtieth of the reserves of the whole country. But then few knew that India had the world's fourth largest reserves of iron ore. Noamundi was a jungle and nearby, in a place called Joda, the adivasis grew rice and maize, ate root and nuts, happily hunted deer, boar and rabbit. Occasionally elephants raided their crops and when the raiders came, they lit fires and threw boulders at them.

In the fourth century BC Chandragupta Maurya had an official whose title was Superintendent of Mines. It was his duty to make borings beneath the surface where ashes or slag indicated workings of mines. Joda was situated in Chandragupta's kingdom and in those distant days, Joda perhaps supplied iron for Maurya weapons of war. Kalinga, scene of the famous battle that changed Emperor Ashoka's life, is not too far away. In 1982, in a Joda village I could see a lean dark-skinned man in loin cloth pushing the bellows with his feet, working for hours to produce a handful of pig iron.

Almost half a century ago when Joda was first discovered by the Tata prospectors, it was ridden with malaria and black-water fever which alone took a toll of up to sixty people out of every thousand per year. A few years later black-water fever was stamped out and malaria mortality reduced to ten per every thousand. How did the local people react to the Tatas coming in and starting mining for iron ore? Charan Patra, a local who has worked thirty-nine years since mining started, says that the local people had 'no reaction' either positive or negative. They took it as a part of life. 'Now,' he says, 'conditions are easier but the jungle is thinner.' And the weather, he says, has changed with the disappearance of the jungles. 'Formerly in winter we had to sit beside the fire. Not so now.' Where once fifty families lived, today two hundred dwell and of these, seventy families have at least one person employed in the mines and some have two.

The mines have transformed the economy of the place. Women carry the ore in baskets. When they first join mining, says an official, they are extremely poor but after the receipt of the first month's salary, they turn up with their first investment: a three-piece tiffin box. Many of them have to walk for miles to come to work. So, after a couple of months, they invest in a bicycle. Riding a bicycle is not quite in their line, so they hire a young man to drive them to and from the mines, the rider having the use of the bicycle the whole day. On these journeys on bicycles made for one, friendships develop and many a couple have taken the path to matrimony through bicycle rides.

It was Jamsetji Tata's foresight to which the company owes its mining rights. As noted earlier, as soon as he got assurance from Lord Hamilton, secretary of state for India, of the government's cooperation, he cabled his Bombay office to buy licences for mining the various minerals required to make steel. It has held his company in good stead even hundred years later.

Joda is one of the eight major mining townships in charge of TISCO. TISCO has its own mines for manganese. It mines chrome at Sukinda, dolomite at Panpush, limestone at Hathibari, fire clay at Belpahar, magnesite at Dodhanya. Each mine is a world in itself with its mine cars and locomotives its hospital and water services; and its electricity, sanitation, housing, canteen, recreation and often a village community development project. Most of the general managers of the mines or divisional managers keep a day in the week when anybody in the area, be it an employee or a citizen of the place, can come and see them and ask for their help on any matter.

The mining towns are a heavenly place far from the madding crowd but if one is there most days of the year, it can get quite lonely. Each of the mines makes an extra effort to set up recreation clubs and to hold sporting events. Basudev Das, the power lifting champion for India in the 1980s, worked at Noamundi and the Tatas sent him to the USA for the world championship. Next only to mining, the greatest responsibility of the management is to make the mines interesting and habitable.

In some areas there are vocational training centres set up, like the one at Sijua Collieries, where every year about 2,000 to 3,000 supervisory staff and workers are trained. Special attention is paid to safety measures.

India has the largest reserves of iron ore next only to Russia, Brazil and Australia. While Brazil and Australia export their iron ore, Russia is not known to do it. When the late Mohan Kumaramangalam, minister for mines, nationalized every coal mine in India in 1974, he left untouched the coal mines of TISCO because, he said, he wanted a yardstick whereby to measure the production of the nationalized coal mines.

The minister requested the Tatas for the services of its

mines manager, R.N. Sharma, as chairman of Coal India, with 350 mines. Sharma had been with TISCO for twenty-five years and when asked what he found distinctive about the company, he said: 'No one expected you to do anything unfair to employees, to customers or your Government. I know of instances when the Railways asked for a demurrage of Rs 5,000. Any private operator or even some well known companies would give a small amount unofficially and get out of it. Tatas would rather pay the Rs 5,000. In the long run, we did not lose really, we gained an enormous amount in respect.'

He continued, 'To us on the staff all authority was given. No one asked why he took on the full responsibility. This approach is conducive to fast growth. We were encouraged to speak out our minds fearlessly no matter who was at the other end. This ability I find most helpful when I am in a Government job now'.

TISCO runs its own geological department and has a senior deputy managing director in charge of raw materials and mines based in Calcutta. When the mines at Noamundi first started, only bachelors were welcome and they had to live in tents. Snakes and tigers had to be reckoned with and they always moved in groups, to and from work. Amanullah who came to work at Noamundi in the 1940s recalled that before they collected their salary packets they had to queue in line and they were each dished out from a large bottle a dose of quinine mixture. First things first! He recalls that once, when his wife fell severely ill, he was asked to report for duty at another mine. He was hesitant to go. The company promised special attention for his wife and an expert doctor was driven all the way from Jamshedpur, 80 kms away to look after her complaint—although Amanullah was a small cog in the wheel. In the 1980s, retired Amanullah told this author with feeling: 'I shall never forget that.' While technology does play its own part in higher production, it is also this kind of personal attention that has made the mine workers of TISCO give their best.

Working at this colliery in the early 1980s was a young and dynamic adivasi TISCO officer called Shalkhan Murmoo, who had deep concern for his people. When Murmoo was working for the rural development society at west Bokaro, he came across a village of twenty-nine families belonging to the most backward of tribes in the whole of Bihar called *Birhors*. These tribals depended entirely on hunting for a living, occasionally making ropes and gathering roots and herbs used for medical purposes. However, hunting was getting more and more difficult as wildlife was disappearing and their own population was rising.

The *Birhors* are the most backward of the adivasis. They live as primitive men must have done before recorded history. The whole family stays in tiny huts with a 3 ft radius, thatched with leaves; to enter one must crawl inside. Murmoo relates:

'When we first started our activities in the different villages of the West Bokaro Colliery, my friends and I went round the area. When I first visited this village of 29 families in my Jeep, they all scattered to their huts or disappeared into the jungles. They were scared. I went back. I returned with a local tribal, a Santhal, and a few villagers gathered courage to come closer to us and were surprised to find I could communicate in their language. They were so happy and excited that one by one they all came out of their huts. Perhaps they sensed that I cared for them. I explained to them our programmes and started a school for children under a tree. We got a teacher to instruct the children in the alphabets for three hours a day. Gradually even the parents began to sit round their children gathering knowledge.

'Soon we got very friendly. I suggested that they take to some work apart from hunting. It was a totally new idea for them. They were a nomadic tribe but gradually the concept of well-digging, gardening and manual work was introduced to them. They needed a more steady income and I suggested that they take to basket-making. Reluctantly they agreed. Our experiment to teach them basket-making with a trainer on the spot was not a success. There were, however, a few young men who were eager to learn. We sent five of them to Jamshedpur at one of the Tata-sponsored centres for basket-making. We provided them all the facilities. They came back to their village and started making cane baskets which were good. Our collieries are now buying the cane baskets which have to be very big in size and strong as they are used for coal carrying. The five boys now make five baskets a day and can earn well. The school has a proper building now, and an adult literacy programme has been started. A kuchha road and a tube-well have come up and we are in the process of persuading them to live in mud houses, so they have the security of living in one place. We have sent four of their children to boarding school for further studies. I am happy we could do something for them in the year we have worked with them.'

11

Vignettes

Jamsetji Tata's birthday, 3 March, is celebrated with each department of TISCO taking out a float in an open truck not unlike the Republic Day parade. It lasts for about three hours. Each departmental float has a theme of what they are doing. For example, digging iron ore or stocking a blast furnace and some of the accompanying floats even have their own music. There is great excitement and the decibel level can occasionally be high. On one such occasion, when this writer was present, a worker jumped out of the float truck as it was passing by the old small bust of Jamsetji Tata. He had obviously come from the last shift in his working clothes. Unhurriedly, unconcerned about the spectators he walked towards the bust and produced a few marigold flowers to lay at its platform. He stood there hands clasped in prayer for quite some time. After that, he turned and melted away in the audience. It was an unforgettable occasion.

Over the span of hundred years, many others have also played a key role in creating a company which according to J.D. Choksi is a 'part of a landscape of India'. Some have appeared in the earlier pages—foreigners like Tutwiler and Keenan, and Indians like Sir Jehangir Ghandy. Some lent their names to the landmarks of Jamshedpur like the Keenan Stadium, Sir Dorabji Tata Park, Sir Ardeshir Dalal Memorial Hospital, Sir Jehangir Ghandy Memorial Hospital and the grand J.R.D. Tata Sports Complex. But there were many, many more who are not commemorated; whose names do not figure and are now a distant memory. Each one of these courageous and interesting personalities has added a line or two in the company's history. While their names may be relegated to the footnotes of history, their work lives on, carried forward by many others.

Prior to 1971, under the Managing Agency system, Tata Sons, and Tata Industries, appointed a 'Director in Charge' of each major company on their behalf. Sir Ardeshir Dalal had that designation for Tata Steel. His father died at a young age and he had to struggle for an education. He stood second in the highly competitive ICS exam. He was the first Indian to be municipal commissioner of Bombay, a post the British had reserved for their own. He joined Tata & Sons as a director in charge of TISCO in 1931. When the company was still going through a difficult period financially and suffering from unstable labour relations, he dared to introduce a profit-sharing scheme

Sir Ardeshir Dalal

for labour in the 1930s. When he passed away, T.P. Sinha of the Tata Workers' Union said that though the union did not always see eye to eye with him, it always 'found him frank and straightforward and a perfect gentleman in his dealings with us, which greatly helped in establishing happy industrial relations'.

A signatory to the Bombay Plan, Sir Ardeshir was invited in 1944 by Lord Wavell, to join the viceroy's executive council as member for planning and development. For fifteen months he applied himself to the task with vigour, laying its foundation. The multipurpose river valley schemes and the chain of national laboratories were first conceived during his leadership of the planning portfolio. He was a precise, firm and domineering administrator. Michael Brown, later editor of the *Illustrated Weekly of India*, described him thus: 'He is rather tall and sparsely built, extremely well-groomed. His heavy lidded eyes and gentle mouth may suggest the dreamer but there is an indefinable atmosphere of preciseness about him. Even his cheroot seems trained only to scatter its ashes in the ashtray.'

When he resigned from the government he returned to his desk in Bombay House in January 1945. He became vice-chairman of Tata Steel and then started planning for the newly formed Tata Engineering & Locomotive Co. Ltd.

The eminent economist Dr John Matthai began life at the bar. He then went to the London School of Economics and got his doctorate there for his thesis on 'Life in Ancient India'. On his return he became professor of economics at Madras University. He came to the attention of the government of India for his accomplishments and was appointed director-general of commercial intelligence and statistics. He was credited with formulating and the actual writing of the Bombay Plan.

In 1946, Nehru invited Dr John Matthai to join his interim government. He became minister of railways and transport, then of commerce and industry and in 1948 became minister for finance. He distinguished himself in all these positions but resigned in 1950 due to differences with Nehru.

'Dr Matthai does not believe in wasting his energy. That appears to be the secret of his success,' wrote a contemporary. But unlike Sir Ardeshir who was volatile in temper, Dr Matthai was a serene personality.

Dr Matthai served with the Tatas for another nine years. When he retired, a colleague wrote of him:

'He respected the dignity of the human personality and did not believe in violating it nor in tolerating any violation. Into the hard pressed and blasé business world he brought an air of learning and love of it too...His vision was without blinkers, and his mind was erudite and enlightened beyond the immediate and the present, soaring high above most things'.

J.D. Choksi came to the Tatas in 1938 as a legal adviser. His finest hour came during the struggle to keep the Steel City of Jamshedpur functioning when the communists declared a flash strike in 1958 to test their strength against the official union. As vice-chairman of Tata Steel, J.D. Choksi rushed from Bombay to Jamshedpur and stood his ground giving confidence to his people. And the official union came out successful.

A towering and impressive figure, J.D. Choksi was equally at home in law, finance, business and administration. He held the reins of one or another Tata company with distinction for three decades. A chance meeting on an aeroplane between J.D. Choksi and George Woods, adviser to the World Bank, resulted in the involvement of the House of Tata with the World Bank. As a result, a loan of US $50 million, the largest ever from the World Bank for Asia till then, assisted the modernization of Tata Steel doubling its capacity. Later Woods became president of the World Bank. The relationship between the World Bank and the Tatas continued gainfully for several years.

Both George and Louise Woods were fond of Indian tea and J.R.D. Tata ensured they had a steady supply. Louise Woods wrote to JRD in February 1965: 'Your tea has arrived per schedule, thank God. I do hope that you have left in your will that I am to continue to get that tea in case you die before I do.'

In his letter of 12 March 1965 JRD replied: 'I have considered your suggestion of making a provision in my will to ensure that you continue to receive that tea after my death, but as I do not expect anything to be left of my estate after paying taxes,[1] I am making other arrangements to ensure continuity of supplies even while I am waiting for you on the other side. I only hope we will both go to the same place!'

It was not just the men that were within TISCO who were precious. There were many that participated in building it from the outside. These were not always well-known leaders or scientists. Some of them were honest and sincere suppliers and workmen whose dedication and sincerity was noteworthy.

In the early 1970s, an enterprising young man, fresh from the Birla Institute of Technology, set up an ancillary factory in Jamshedpur hoping that Tata Steel would give him business. He started with small orders running from Rs 2 or 3 lakhs (equivalent to Rs 30 to 40 lakh today). When the head of the ancillary department, Mr Hassan, called him for drinks, the young man told Mr Hassan that he found it embarrassing that a small supplier like him was being invited by the head of a department.

Mr Hassan smiled and explained: 'I make inquiries about the people whom we deal with. I know that you are dedicated to your work, very hard working and honest. You don't use underhand means to get business from us. There are others who corrupt our officials to get business deals. We worry about our administration getting corrupt. Therefore, we encourage businessmen like you so that ultimately we stand to gain. It will then be possible to stop dealing with those who use corrupt practices.'[2]

---

[1]In those days, income tax laws taxed income in the higher brackets as high as 95 per cent.

[2]*My Journey*, by Pratap Pawar, *Sakal*, Pune.

The young man went on in the years to become the managing editor and trustee of one of the most respected Marathi papers published from Pune, *Sakal*. He has grown both in industry, in journalism and in the social life of Pune. His name is Pratap Pawar.

For forty-five years, TISCO was the most significant steel maker in India. When Nehru decided on government steel plants, he was thinking of the scale of 5,00,000 tonnes. J.R.D. Tata told this writer he advised the prime minister if the government is moving in, it should think in terms of 5 million tonnes, not 5,00,000 tonnes.

When the three plants of the Steel Authority of India were started they had little or no indigenous manpower to draw upon compared to the kind of expertise that the Tatas had. And so understandably they drew heavily on TISCO's manpower. Fortunately, such was the manpower strength of TISCO that its progress was unaffected by this outflow of talent. Russi Billimoria, its chairman, was also a top TISCO official.

TISCO also lent its trained executives to the Tata organization for their other needs. Tata's New Delhi office often drew on the talent of Tata Steel for its chief executives like K.C. Mehra and Sujit Gupta.

In 1956, Syamal Gupta, a twenty-one-year-old graduate engineer from Jadavpur University joined TISCO. He joined at an exciting time when, with the help of Kaiser Engineers from America, the capacity of Tata Steel was doubled from one to two million tonnes. Recalling those times half a century later, Syamal as Director of Tata Sons, speaks fondly of his superiors, Savak Nanavati, Vishwanathan, Homi Bodhanvala and Firoze Tarapore. 'These men,' he says, 'nurtured us.' His remarks are indicative of the strong culture of mentorship and nurture that existed at the Tata's. In those days people in the corporate world were not so busy climbing up the ladder as to neglect those struggling on the lower rungs. 'There was a thrill as the new Smelting Shop came up—and there was friendship and fellowship.'

A German in charge of the rolling mills perceived the latent potential of Syamal and arranged for him to spend a year in Germany to train. From Germany, eager to learn further, Syamal, on his own steam, went to the Imperial College, London. Though he had gone on his own, senior officials persuaded management to adequately reimburse the young man—and they did. Meanwhile, having refused other offers abroad Syamal rejoined as design engineer in the Growth Shop of TISCO.

Unknown to him, J.R.D. Tata, Sumant Moolgaokar and Freddie Mehta, Tata Directors, had a meeting with the prime minister of Singapore, Lee Kuan Yew, in 1971. Singapore then was known for fishing and garments. Lee's view was that Singapore could industrialize and asked Tata who were only in heavy engineering, to start a Precision Engineering industry plus a training institute for his young people. It was a tall order, for India was also a closed economy short of foreign exchange. But they took up the challenge to manufacture semiconductors and computer peripherals. The Tatas had to train themselves first. They chose Syamal to start the project. 'In those days, there were no websites to refer to,' Syamal adds, 'but Siemens did help.' Not long after, GE, Siemens and other international giants became customers.

Ten years later, Syamal's former colleague in Jamshedpur, Ratan Tata, posed Syamal the question, 'What about Africa? Have you ever thought of it?' Syamal had not, but he took off on a tour beginning with Africa's best known sight, Victoria Falls. To his utter surprise, he found a steel beam there which had on it, the words 'Tata Iron & Steel Co Ltd, 1933'.

And so the discovery of Africa for the Tatas began through a TISCO man. His initial thrust was with Telco (Tata Motors) trucks. They were particularly well-suited for the rough African roads. Tata interests have now broadened into other fields. It is not accidental that Tata Steel's first overseas location in 2006 was at Richard's Bay in South Africa, where it set up a ferro-chrome plant. South Africa is one of the largest producers of ferro-chrome, an essential alloy for both stainless and carbon steel, in the world.

The Tatas have established themselves in over ten countries of Africa. Ratan Tata is on the India–South Africa CEOs Forum as Co-Chairman. Syamal is on the Advisory Council for Business on Investment for Ghana, Tanzania, Zambia and Uganda (the first and last two are World Bank appointments).

Part IV

Touching the Lives of People

12

Citadel of Sport

Tatas represent the spirit of adventure.

—Mahatma Gandhi

Bachendri Pal, the first woman to be on top of the Everest, was almost blinded by the snowstorm during the last stretch. Strong-willed, she strived and reached the top. There she hoisted, along with the Indian and Nepalese flags, a flag that belonged to Tata Steel. She has been in charge of many of the Tatas' adventure facilities like mountaineering, trekking and the like for several years. When this writer met her, he found that she came from a high mountainous area in Himachal Pradesh. And in that area, where oxygen is rare, they develop a lung power which we, who live in the plains, do not. This advantage, plus a hardy life, also enabled her to achieve this feat.

The Tata Steel Adventure Foundation has more than 3,000 people enrolled every year. They are not only younger Tata trainees but executives who join in a two-week mountaineering course at Uttaranchal. Sports develops discipline, team spirit and endurance. It reflects in their work later.

J.R.D. Tata was seventy-eight when he flew from Karachi to Bombay, to commemorate the first flight in 1932. Speaking to a galaxy of personalities, JRD described the flight he had just undertaken in a single-engine Leopard Moth of the 1930s with no aids other than a speedometer and an altimeter. Even a map of the ground was rolled and kept between his legs as on the first flight of 1932. The only difference was, he noted, there was a radio installed which was crackling away and indicated it was more of a nuisance than a help. He added he was surprised nobody asked him why he re-enacted this flight at seventy-eight and went on to answer his own question:

'This flight of mine is intended to inspire a little hope and enthusiasm in the younger people of our country that despite all the difficulties, all the frustrations, there is a joy in having done something as well as you could and better than others thought you could.'

Jamsetji Tata had written from the US to his son, Dorab, about his dream of a steel city with shady trees. He had specially asked: 'Be sure that there is plenty of space for lawns and gardens. Reserve large areas for football, hockey and parks...'

And so it all happened. Sir Dorabji Tata, who implemented his father's dreams, was himself a sturdy sportsman who rode on horseback from Bombay to Pune in eighteen hours. He sponsored the Indian contingent to the first two Olympiads from his own resources; found India a place in the sports world. He was succeeded briefly as chairman of Tata Sons and Tata Steel (1932–38) by Sir Nowroji Saklatvala who was passionately fond of cricket and, when the Cricket Club of India (CCI) came up in the 1930s, sponsored the JN Tata Pavilion—the centerpiece of the CCI.

J.R.D. Tata himself played tennis and started skiing at the age of forty when most people retire from this exacting sport. He went on to ski for another forty years. Ratan Tata wanted to start a flying club in Jamshedpur and this brought him and J.R.D. Tata closer to each other.

The sports infrastructure in Jamshedpur is second to none in India and half the users of these facilities are also citizens of Jamshedpur, unconnected with the Tatas. In addition to the Keenan Stadium, there is a magnificent JRD Tata Sports Complex with a seating capacity of 40,000 and with an eight-lane synthetic athletic track. In the complex are two basketball courts, tennis courts and a hockey ground. Volleyball, chess and boxing are also available.

Horse riding, polo, swimming, golf and aviation are some of the other facilities available. What is most encouraging is that surrounded by an adivasi habitation, the company took interest in archery, which is the original sport of the adivasis. The sport used to come handy for their food needs. Training in archery is quite expensive. The most modern equipment was made available for the Tatas Academy for Archery in 1996 at a time when the company was still battling to survive.

The Tata Archery Academy, inaugurated in 1996, was aimed for the revival of an ancient indigenous sport of a people on whose land the plant was built. Not only adivasis of Jharkhand, but others from Assam, Nagaland, Mizoram and Manipur have also got innate skills for it. Purnima Mahato, winner of seven gold medals at the National Archery Championship and an assistant coach since 2002 says, 'These Archers would not have been able to pay for equipment in ordinary circumstances. They also get exposed to international trends and standards here.'

The Tatas' policy on sports is two fold. One, is to take promising people in their twenties who have shown talent and give them all the facilities they need. Second, is to take youngsters as young as fourteen or fifteen and train them in an institution like the Tata Football Academy (TFA), which was started in 1987. For four years, all their needs are looked after. They are given education to pass their main examinations and they also get a stipend. This is a long-term investment in the sport. Football, once very popular in eastern India, is now catching on elsewhere too.

The trainers at the TFA were trained in Sao Paulo in Brazil. Today, from the TFA at least twelve of the boys have emerged with high Asian standards of play. Out of 129 cadets who have been trained, 105 have represented the country and seventeen had the distinction of captaining the national side.

Spurred by the success of Tata Football Academy and the Tata Archery Academy, the Tata Athletics Academy was inaugurated in the presence of Flying Sikh Milkha Singh, in May 2004. In a span of three years, several cadets of the Academy have won medals at various National Meets in the Junior and Senior categories. Ms Sinimole Paulose won one gold and one silver medal in the Asian Indoor Athletic Championship 2006. She also won a silver medal in the SAF Games held at Colombo in 2006.

Providing necessary training facilities—especially to the young—enables them to excel in their respective disciplines and win medals at the international and national levels through training centres for athletics, basketball, boxing, badminton, chess, cricket, golf, handball and tennis.

School sports are encouraged and courses on fitness, exercise, yoga, meditation and nutrition are held on a regular basis, which helps in ensuring a fitter work force that can overcome the stress and strains of the workplace. These courses are open for all citizens of Jamshedpur.

As sports for children with disabilities are often neglected, various sporting activities are conducted under the banner of Special Olympics, Jharkhand, for physically and mentally challenged children to integrate them into the mainstream of society. Annual sports programmes are organized in football, athletics, floor hockey, basketball, badminton etc. Tata Steel had also sponsored the Indian team in the Special

Olympics which were held at Dublin, Ireland, in July 2003. Three children from the steel city had performed outstandingly and won three medals.

International and national matches have been played for various sports, the more recent ones being:

Cricket

ODI Cricket Match – India vs England	1992-93
ODI Cricket Match – India vs New Zealand	November 1995
Three days Test match – India vs Australia	March 1997
ODI Cricket match – West Indies vs India	November 2002
ODI Cricket match – India vs Pakistan	April 2005

Archery

Sr National Archery Championship	March 1992
National Archery Championship	December 2005

Golf

Tata Open Golf Championship	November 2004
Tata Open Golf Championship	December 2005

Others

National Athletic Championship	September 2003
JRD Tata International Invitational Badminton Championship	September 2003
Tata International Open Chess Championship	October 2003

The incredible part of the Tatas' approach is that even when struggling for survival in the late 1980s/early 1990s, it not only kept sports in the frontline, but started new institutions like the Archery Academy, and sent the archers abroad for international competitions. Sports are not a luxury but a path to create a complete personality.

The present managing director of Tata Steel, B. Muthuraman, is an avid golfer. In an article on golf, he wrote what it means to him. His golf foursome has top-notch players. He writes: 'Our four-ball is unique. There are two champion golfers—C.D. Singh and Ravi Sharma. The third one, Hindi Grewal is an ex-champion. He is "all in one" for us—still a solid 12 handicapper, coach, guru, sensei and the expert. The fourth is the undersigned. Play-

B. Muthuraman

ing in the four-ball is serious stuff. The only one who is allowed to speak is of course, our Coach-cum-Guru. The three of us listen to him intently. But we play our own game. We have successfully learnt to disconnect the advice being persistently given by our Coach and the way we play our game. While being a part of our four-ball is serious stuff, it is also a great joy. Some of the happiest moments of my life have been on the golf course. Long after I leave this town and as I get old and my memory fades, the last thing that will remain with me, apart of course from my wife and children, are the great moments that I have had playing in my four-ball and the tremendous contribution the three of them made to enrich my life.'

13

Social Conscience of Industrial India

What happens in and to this community vitally affects us all, and that is why Tata Steel is known throughout the length and breadth of the country, not just because of the steel it produces, but because it is also the social conscience of industrial India, interested in every facet of life of all its peoples.

—J.R.D. Tata

In the leisurely 1930s, during the hot noon hours, most officers used to take a break and stretch out under a fan reading or resting. The general manager, John Keenan, was doing this one afternoon in 1934 when:

'Suddenly my wife landed on the floor. She claimed that in an access of playfulness I had pushed her off the bed. Then we both noticed the queer stillness of the air. We saw that the large overhead fan was swinging crazily in all directions. There was a deep rumble beneath us. We rushed into the garden. It was an earthquake.'

At nine that night, a telegram came to TISCO from the governor of Bihar and Orissa, Sir James Sifton:

NORTHERN BIHAR AND ENTIRE GANGES VALLEY DEVASTATED BY EARTHQUAKE STOP CAN YOU HELP WITH DOCTORS NURSES MEDICAL SUPPLIES CLOTHING GALVANIZED SHEETS STOP LIGHT STRUCTURAL SHEETS FOR TEMPORARY HOSPITALS MEN STOP MOST URGENT STOP

The Bengal–Nagpur railway put a special train at the disposal of the Tatas. In Jamshedpur, men worked the whole night. At eight the next morning, the relief team steamed out of Tatanagar with fourteen doctors, six nurses, a cartload of blankets, bandages, medicine and a complete field operating theatre. Other freight cars carried 4,000 tonnes of galvanized sheets from Tata stores, beams, nuts, bolts, rivets, angles and 200 tonnes of rice. A dozen engineers, 100 mechanics and the Tatas' Welfare superintendent, K.A.D. Naoroji, set off on a circuitous route for Monghyr, the epicenter of the earthquake. Rescue workers steered round huge crevices.

Within a day of the arrival of the train, a temporary hospital had been set up for 400 patients. Thousands were being inoculated against cholera, smallpox and tetanus. When the viceroy, Lord Willingdon, arrived by plane from Delhi, the Tata's relief staff was just getting underway. Monghyr was in ruins, but there, as if by magic, stood a functioning hospital on one side and on the other, a sheltered enclosure to feed the homeless. Fifteen thousand people lost their lives in the earthquake. Tata teams helped to extricate from the ruins the living and the dead. Thousands were saved. Lord Willingdon sent a cable to Keenan:

TATAS HAVE DONE SPLENDIDLY—WILLINGDON

Three of the Tata officials were decorated in the king's birthday list but, says Keenan: 'more important, all India suddenly realised anew the value of the Company in times of need, that indeed Tatas were not out to exploit the country but to benefit it'. The grumbling of the shareholders (who were not getting any dividends at that time) was replaced by a rightful pride in their company.

Since those days in the mid-1930s, the company has helped in every national disaster. During the Patna Floods of 1975, the Tata relief plane was the first to land and TISCO workers voluntarily donated one day's salary to help the citizens of the city. The cyclone which devastated part of Andhra Pradesh and the earthquake that shook Koyna received similar assistance. TISCO's relief volunteers have now become highly proficient at their work and in fact the company knows it can depend on a reservoir of 'disaster-expertise'. Be it during the Uttaranchal or the Bhuj earthquakes, the Tata Relief Committee at Jamshedpur is one of the first to move in. It rescues and it builds anew for the future. Usually the management matches the total contribution of the staff and workers which may be one or two day's salary.

A LINE IN THE HISTORY OF MAN

TISCO has had a part in the eradication of smallpox from the world.

In 1974, two Japanese were found to have contacted smallpox and it was traced back to Tatanagar station. WHO, which had earlier launched a campaign to eradicate smallpox for ever, was greatly concerned. They rushed off Dr Larry Brilliant, an expert, to Tatanagar where he called on the managing director, Russi Mody. Without batting an eyelid, Dr Brilliant asked him for fifty doctors, 200 paramedical supervisors, upto 900 vaccinators-cum-researchers, and fifty vehicles. When Mody asked when he would like these facilities, Brilliant replied: 'From yesterday.'

The response was instantaneous. The Tatas put thirty-two jeeps and other vehicles at the disposal of the WHO team, while OXFAM and WHO produced another thirty-six. Within seventy-two hours, staff were recruited for vaccination, trained and put on the road. They scoured a 45-mile radius of Jamshedpur. Lakhs were treated and vaccinated. The name 'Tata' acted like magic in places where the tribals were otherwise suspicious of outsiders. There were, however, occasional hazards as when a vaccinator, who knocked on the door of a tribal hut, found two ladies charge at him with darts and arrows. The vaccinator ran for his life.

It was not Tatanagar alone but the whole of Chhotanagpur that was afflicted with smallpox. The Tata Steel board met and sanctioned a major project in which Telco (now Tata Motors) and other Jamshedpur-based companies participated. The first phase was completed as scheduled in six months. The second phase was that of containment when Tata officers had to virtually risk their lives by going into the most unhygienic surroundings in the area often in remote jungles. Operation Containment was declared highly successful and what is more, Rs 10 lakh were saved from Tatas' allotment because of strict expenditure control. WHO requested the Tatas to join in Phase 3—of Consolidation. If this phase was successful, the dream of WHO to eradicate the menace could be achieved. The government, WHO and Tata teams maintained surveillance from February to June 1975. One day, a message came from WHO, New Delhi, to Jamshedpur and Bombay—the battle against smallpox had finally been won.

'Perhaps,' said Russi Mody, 'by the efforts of those involved in the campaign, a line in the history of mankind has been written.'

HUT HOSPITAL TO RESEARCH CENTRE

When the hospital began in 1908, it was a small tent about 80 feet square, partitioned into two small sections, one a dispensary and the other the chief medical officer's room. In the dispensary was one small almirah with some phials and dressings. The chief medical officer was Dr Shantiram Chakravarti, who went on horseback on his visits. P.C. Mitra recalls that a few days after the opening of the hut hospital, he saw one morning a man suffering from herpes, standing by the road. When Dr Chakravarti and he spoke to the man to come in and get treated, the man shouted to others hidden in the jungle and one by one, seventeen men came out of the bushes—all suffering from the same disease!

The majority of the workers were tribals who were sunk in abysmal ignorance and superstition, and were deeply suspicious of the hospital and modern medicine. The doctors, therefore, not only treated the patients, but also taught them with missionary zeal the elementary rules of hygiene and health care.

The hospital has grown over the years, and the Tata Main Hospital is one of the foremost in the country with 900 beds, fourteen specialized departments and 168 doctors. Here, the humblest worker can get the attention of the same specialist as the managing director.

Facilities are available without payment to every single employee of Tata Steel and apart from the staff and their spouse, the children are looked after till the age of twenty-one. Also, the elderly parents are cared for.

Wherever Tata Steel has an industrial base, it has reached out its hand and expertise to others. Furthering its partnership with the Jharkhand government and its people, Tata Steel is committed to contribute Rs 25 crore every year, for the next thirty years, to a government-run health insurance scheme for the state's Below Poverty Line (BPL) families who would avail a medical insurance umbrella at zero cost.

The company's involvement in the sustained development of Orissa, where most of its mines are located, continued over the financial year through its new education projects. It set up the JN Tata Technical Education Centre to impart technical skills among the youth of the state so as to improve their employability in industry. Tata Steel, in collaboration with the state government, also launched a project for setting up the Institute of Mathematics and Applications to promote the culture of mathematical competence at all levels of society

To honour the first Indian general manager, and later managing director, Sir Jehangir Ghandy, a block was named Jehangir Ghandy Memorial Hospital. On Founder's Day in 1989, a fierce fire lit the grandstand area. The hospital facilities were a help but inadequate to treat severe burns. TISCO, thus, set up a sophisticated 'burns' unit and a specialized unit for the reconstructive surgery of mutilated hands have been added plus a Critical Care Unit apart from the ICU. The Rehabilitation Centre also will be modernized. They have had years ago a well-designed ICU. It is worth mentioning that J.R.D. Tata was at Geneva when the fire broke out. He flew to Jamshedpur early morning and visited all the bereaved families.

In addition to the Tata Main Hospital Complex, Jamshedpur is proud of one of the finest, though small, cancer hospitals, the Lady Meherbai Tata Hospital, also recognized as a research centre by the All India Medical Council. One of its radiotherapists, Dr K.D. Reddy, was chairman of the 13th International Cancer Congress at Washington, 1982. In addition, there is the Sir Ardeshir Dalal[1] Hospital for tuberculosis. The hospitals of Jamshedpur serve the entire region, irrespective of whether a patient is a Tata employee or not.

---

[1]He was Tata Sons Director in charge of TISCO.

14

Reaching the Unreached

*I do believe that we, in the Tata Group, have held
a view and a sense of purpose that our companies
are not in existence just to run our business and to
make profits but that we are responsible citizens
over and above our normal operations.*

—Ratan Tata,
Chairman, Tata Steel, 1993–till date

By the end of the 1960s, the Tatas felt that they had
done all that they could for the staff, the workers and
their families. Possibly more than had been done by any
other company. However, Jamshedpur was an island of
excellence in a sea of poverty. Speaking at the Anantha
Ramkrishnan memorial Lecture in Madras, in December
1969, J.R.D. Tata said:

'Every company has a special continuing
responsibility towards the people of the area in
which it is located and in which its employees and

their families live. In every city, town or village, large or small, there is always need for improvement. Let industry established in the countryside "adopt" the villages in its neighbourhood; let some of the time of its managers, its engineers, doctors and skilled specialists be spared to help and advise the people of the villages and to supervise new developments undertaken by cooperative effort between them and the company. Assistance in family planning in the villages would be particularly valuable form of service. None or little of this need be considered as charity.'

Words were followed by speedy action. The Articles of Association of Tata Steel and some other Tata companies were altered with the shareholders permission so that resources could be utilized for the benefit of those beyond the confines of the factory. Christopher Dawson wrote, 'Behind every civilization, lies a vision.' With JRD's vision, a civilization of extended care for the needy was born.

The pink lotus sits gently upon its fan-shaped green leaf amidst the still waters. Spread around it, is the Jubilee Park given to the city of Jamshedpur by the company in 1958 on the occasion of its Golden Jubilee. When the then prime minister, Jawaharlal Nehru, came to inaugurate it, he looked around and said:

'Flowers, parks and houses supply something which is, I imagine, of more basic importance to human beings and the human spirit than even iron and steel, and it was a very happy thought to commemorate the occasion of the Jubilee of this great company by providing this beautiful park.'

Nehru unveiled the statue of Jamsetji Tata. The statue faces the bluish-green Dalma hills where, even today, a lone tusker is quite capable of squatting on the narrow road going to the hill top, halting the journey of any VIP. One

Cadets at the Tata Archery Academy.

Cadets at the Tata Football Academy.

Tigers in the Tata Steel Zoological Park, Jamshedpur.

Tribal dancers from around Jamshedpur.

can neither navigate round the pachyderm nor persuade it to give way. Behind Jamsetji's statue in the distance is a landscape of the entire steel plant. Under Jamsetji's statue is the Latin inscription which says:

'If you want to see his monument
Look around you.'

Not too far from the statue are flowers of a different kind in a place called Sishu Niketan. Here, since the 1950s, members of the All-India Women's Conference (AIWC), Jamshedpur Branch, run a home where children of patients suffering from leprosy grow up as healthy human beings. Blossoms that may have been lost in the dust, flourish in the soil that gives them thought, friendship and care. The patients who leave their children behind know that in doing so they surrender their claim on the child and should keep no contact with their progeny. But sometimes, even years later, they come unobtrusively and watch from a distance as their offspring play in the gardens or go to school. What strikes a visitor about this home is the extent of involvement of the ladies who run it, be it the Convenor, the matron or members of its committee. Sitama, a young widow, is the matron, who teaches the girls housekeeping, cooking and care of children.

While many boys are adopted quickly, it is usually a slower process for girls. Wives of senior officials who run the home carefully select husbands for the girls and even arrange for ornaments for their wedding. Local shops are requested to donate the clothes and other accessories that parents normally give to a daughter at her wedding. Parting is often difficult, for some of them have spent eighteen or twenty years, and the matron, at times, weeps as she waves them farewell.

The Family Welfare Department at Tata Steel was set up in 1950. It was converted to a trust in 2000. Today, it runs twenty-one family planning clinics.

Young widows of workers, destitute women and

handicapped men are given work at the Kalyan Niketan, which produces working gloves, uniforms, stationery and the like for TISCO, Tata Motors, Indian Tube and other industrial establishments. These workers enjoy a provident fund, medical allowance and other benefits. The profits of Kalyan Niketan are funnelled to finance Sishu Niketan with its thirty-four children. 'We would never be able to do this but for the help of TISCO', said Mrs Homi Bodhanvala, who was president of the AIWC in Jamshedpur. Mrs Bodhanvala's husband, Homi did remarkable relief work at times of drought, floods and communal disturbances. It is people like this couple who bring to bear upon the steel city the steady glow of a social conscience, without which the material prosperity of Jamshedpur could have made it a soulless city. They inspire many others to tread this path.

To start the steel plant, Jamsetji selected the finest consulting engineers of the world. Five years after the plant started, his sons selected the most outstanding social scientists of that period to come and make suggestions for the welfare of the workers. The advisers, mostly connected with the London School of Economics, consisted of Sidney and Beatrice Webb, well-known sociologists, and professors L.T. Hobhouse and Urwick of London University. They chalked out a programme to be executed along scientific lines. For the execution of this programme, the Tatas requested the Servants of India Society to spare one of their ablest social workers. They selected forty-year-old engineer Amritlal Thakkar (later known as Thakkar Bapa). Prices of essential commodities like food grains had soared as a result of the war. When Thakkar Bapa joined in August 1918, his first task was to open fair price shops, for the company's 25,000 employees, at a cost of ten lakh rupees. Having done that successfully and by eliminating the middle-men he turned his attention to opening cloth shops. The company made direct purchases from mills and managed to supply cloth to their workers at one rupee per yard when it was one rupee fifty paise before.

Indebtedness was the main problem of the labourers and Thakkar Bapa organized a dozen cooperative credit societies for different categories of workers which helped to pay off the Pathans who extracted interest rates that ranged to 150 per cent of the amount borrowed. All this Thakkar Bapa accomplished within six months. For the next half of the year he presented to the company his plans for the primary schools, sports clubs, children parks, canteens and airy open quarters for the entire labour force. The company accepted his proposals and set to work. In the years to come there were many demands on Thakkar Bapa from elsewhere in India. He had laid a solid and reliable foundation on which future generations in Jamshepdur could build upon.

'IF EVER I GET BACK MY SIGHT'

In 1958, TISCO established its first community development centre. Today, there are nine of them in Jamshedpur. Its dual purpose is to teach crafts to the destitute, the widows, the unemployed and the young; to provide recreation and to make available a place which is readily accessible for any social assistance they may need. Even social issues such as dowry are tackled and parties opposed to the system are brought together.

The community development centres encourage activities like music and dance and lay emphasis on health education for the community. However, their main thrust is to give training for a livelihood to those who have none, and enrich the quality of life for those who are employed. Women who are in need of help, especially young widows, are taught sewing, embroidery, cooking, book binding, doll-making and teaching. Training in midwifery is also given. Young girls of marriagcable age are also taught these skills. Young men are taught typing, air conditioning, refrigeration, plumbing, motor mechanics, carpentry and the like suitable for them.

Every evening, the community centre is a hive of activity with sporting events like basketball and other such games. One handicapped person who has found new life is Suniti Bose (nee Dutt) now in her thirties. Sister of a TISCO employee at a colliery, she was blinded by an illness as a child. TISCO gave her a stipend to study Braille at the Dehra Dun centre for training of the blind. After the training she was determined to be independent and self-reliant. The community centre at Sonari gave her the task of stitching buttons on uniforms of TISCO staff. One day a man called Kishan Narayan Bose came along and offered to marry her. The well-off ladies of TISCO decided that she would have a marriage as good as most; so they collected money, got shopkeepers to donate saris and even some silver jewellery. Suniti was married in style.

Suniti found the security of a job, a sense of her usefulness and a partner in her life. She has a passion to be normal. She is independent, manages her home and even does her own cooking. When this writer met the couple, in the 1980s, the husband, a bit older than her, insisted that he would like to give her one of his eyes if a transplant is possible. She vigorously disagreed. 'What if it does not work out? You will lose one of yours,' said Suniti, pragmatically. But the hope that her sight will return abides in her. 'If ever I get my sight back, I will be able to cross the road on my own,' she softly says.

SOCIAL AUDIT

Tata Steel assists hundreds of villages in the neighbourhood of Jamshedpur and of its far-flung mines of coal and iron ore. The village projects include agriculture, dairy and poultry farming, piggery, cottage industries horticulture and forestry. Tata Steel is the first Indian company to have a social audit of its activities by a small group of personalities whose judgement and views command nationwide respect and confidence. The committee consisted of the former chief justice of Bombay, S.P. Kotwal, as chairman and professors P.G. Mavalankar and Rajni Kothari as members.

MAKING A MARRIAGE WORK

Physical pain is not the only suffering mankind has. Most of us suffer in a far greater degree from mental pain and tension. Most family counsellors and social workers find themselves dealing with it extensively.

A man and a lady worker from the Employees' Welfare Service called unexpectedly and asked if I would be willing to join them in their counselling service. They said that two couples, related to each other, were being counselled. A brother and a sister of Bengali origin from Jamshedpur had married a sister and a brother from Gujarat. They both had sons, about two years of age. The unhappiness of one couple had created disharmony for the other. As I entered the room, there was a pall of gloom. Sensitively, a lady social worker drew them out one of by one to speak their minds freely. They did. The lady who spoke the least wore a red sari, had sharp features and a dark complexion. Her face was a study in unhappiness. Stone-faced, she sat through most of the discussion. It seemed that the daughter-in-law at Jamshedpur had walked out of the house in a huff and lodged a complaint for harassment and maltreatment against the father-in-law and the husband. The father-in-law, in turn, had lodged a police complaint accusing the daughter-in-law of carrying away her marriage jewellery without permission. The two couples sat quietly for a while, introspecting to see if any of them were even 10 per cent wrong, even if the other may be 90 per cent wrong. They spoke in turn what they were beginning to discover about themselves. The counselling made them face the truth about themselves.

After about one and a half hours, I left. The session went on for another four hours. Some days later, I met the social worker who said: 'You will be happy to learn that the Jamshedpur couple has agreed to reconcile and to withdraw their police complaints. They want to make a go of their marriage and to celebrate it with a tea party at which they want you to be present.' I went to the family tea party in a small crowded room in the *basti*. There was an air of relaxation, both the wives looked at peace, as each held her little son. The glum lady in red was unrecognizable. 'What has happened to you?' I teased her. She replied with a shy smile. Her countenance was radiant.

Part V

Struggle and Triumph

15

The Crisis Finds the Man
(1980–1990)

Indira Gandhi, the then prime minister, came for the Platinum Jubilee celebrations in 1982, as her father had for the Golden Jubilee in 1957. She was accompanied by Rajiv Gandhi. The city was beautifully decorated. Coloured lights were playing at fountains and the entire town was lit up, creating the ambience of a fairyland. An exhibition of photographs and handicrafts made by the adivasis of the region was displayed. After visiting that, the prime minister accompanied by J.R.D. Tata, came on the dais, within the works. 'There is something special about Jamshedpur, about Tata Steel, about its workers about its management,' remarked JRD. The prime minister noted: 'Jamshedpur occupies pride of place in the history of India as it has pioneered steel making in the country. I congratulate those who are working here for their good work and also for the spirit of friendship and amity which exists, and helps resolve differences peacefully.' As the prime minister's cavalcade drove through Jubilee Park, floodlights of different hues illumined the trees.

It was a company that the country could be proud of. It had not only pioneered the steel industry, but from its inception it tried to establish remarkable facilities for its workers and staff, and in the process achieved many firsts in the field of labour welfare. Not only that, it had later extended its welfare programmes to communities living in the surrounding areas as well.

Indira Gandhi and Rajiv Gandhi being conducted around the exhibition by J.R.D. Tata.

Few were aware of the cloud that loomed over the horizon as the company entered the 1980s. Three decades of government control of its price had permitted the company to breathe and tick along and give its shareholders some revenue. But this had left the company with virtually nothing—no reserves for replenishing its plant and machinery. The dictates of socialism had throttled Tata Steel. Two earlier attempts at modernization had failed. In one instance, a day before the board met to agree to a plan worked out by Atkins in England, the rupee was devalued. Costs would have shot up by over 30 per cent had the plan been carried out and thus the modernization was shelved. Thereafter, whenever company officials recommended modernization, it was done in little bits and pieces. Some of the plant's machinery dated to 1920s and some to the 1950s. Money was short, and to raise it in the share market was not easy even in the early 1980s. If the government gave a loan it reserved the right to convert it into equity.

Finally, in desperation, metallurgist Dr J.J. Irani told the chairman, J.R.D. Tata: 'If we do not renovate the plant totally, ten years from now you and I will be standing outside the gates selling tickets to people to come and see the steel museum.'

JRD laughed, but took it seriously and firmly said: 'Do not listen to what these finance people say. You send me your plan.' That was the decisive moment for Tata Steel. He could do what few other men could. He got the board to sanction a complete modernization plan in four phases over the years. You cannot turn around a steel plant in a year or two. And steel plants require more than one enormous machine. Nor is one latest plant component adequate. As Dr Irani explains: 'You cannot make a chapatti with a golden rolling pin. The ingredients have to be looked at first: that they are of the best quality.' The captive coal and iron ore which Tata Steel had was not always suitable for the plant and some had to be imported. The wastage had to be checked. For a decade the research team worked on ways of making local coking coal suitable and finally it succeeded. The ground work was being laid.

A fortuitous incident took place in 1988. Operating heads of fifteen Indian companies were invited to Japan to study their value engineering which had enormously benefited Japanese industry and put Japan in the front rank of the industrial nations within a short time. Dr Irani was among them. On his return, he sent a team from the company to Japan to study their methods and processes. Tata Steel decided that the second step along with modernization of the plant was to inject the culture of quality into its products if it was to withstand international competition. This was a decisive moment in the modernization for not only did the plant have new equipment, but a highly quality-conscious culture was being introduced within its workforce.

The company was still operating on outdated open hearth furnaces of the 1920s. This was the first to be replaced with a modern Linz-Donavitz process beginning 1988. That was the start of physical modernization. Ordinarily, it would have been easy and cheaper for the plant manufacturer to send out to India a few experts to teach TISCO about the new equipment. However, Dr Irani insisted that Tata Steel sends its best people to Britain so that they get complete exposure to the technology and equipment being brought into Tata Steel. The board approved and 120 of the best people were sent.

The first phase was completed before liberalization in 1991. But three vital phases remained. They followed with speed and by 2000 a brand new plant came up at Tata Steel. The phases were:

Phase I   :   Setting up of new LD process steel melting shop and billet caster
Phase II :   Sinter Plant and Bar & Rod Mill
Phase III :   LD Shop II and Slab Casting Machine for plates, sheets; a new blast furnace
Phase IV :   Hot Strip Mill—making coils of steel sheet and later a Cold Rolling Mill

Of all these items, the most dramatic acquisition was the 'G' Blast Furnace. Russi Mody, when on holiday in the Mediterranean, came to know of a new blast furnace, which Portugal had ordered, but decided not to instal because the country wanted to join the European Union (EU). The EU, at that time, had said that there should not be any addition in steel-making capacity in the EU countries. Sacrificing the blast furnace was a small price to pay for Portugal to enter the EU. Russi Mody sent word to his office in Jamshedpur that if they wanted it, they should come quickly and see it. Mr Srivastava of TISCO flew promptly to inspect and Dr Irani followed not long after. There was hard bargaining. Portugal knew they were in a weak position because customers for a huge blast furnace were not common. Finally when the shipment arrived in Orissa, the customs could not believe the price was so low. They released the shipment but levied a substantial fine. TISCO took the matter to the court. The case is still pending.

Dr Amit Chatterjee, the then chief technical officer of Tata Steel, says, 'The advent of the G Blast Furnace virtually at a throw-away price brought in a totally new era of blast furnace iron-making at Tata Steel in the process the company saved the 3 to 4 years it would take to engineer and build a blast furnace.'

However, all this was not rushed. An old company takes time to move and accept change. Change is never easy to implement or to accept. Therefore, all good practices were not taken up at once. It was one after the other. The change was steady and incremental. There were no flashy moves but a careful formulated strategy, followed by effective communication born out of transparency of the management and the trust of the workers.

After that, a fifth phase was introduced. This consisted of skill enhancement programmes for employees. This intended to raise their capabilities to match the requirements of the job. It was appropriately called 'Modernisation of the Mind'.

The ISO-9000 certification came in Tata Steel before it was known nationally and soon after the Six Sigma programme was introduced. Later, to upgrade the quality of all the Tata companies, a model of Malcolm Baldrige awards, instituted in USA, was adopted under the banner of JRD QV. One after the other, the Tata companies turned quality conscious and improved. However, Tata Steel remained at the top. As one machine after the other was being imported, it became clear that to withstand international competition, there would be a need for rationalization of the number of staff.

It was the most difficult bullet to bite. While other efforts were going on, suddenly on 17 January 1991, the government lifted the control on prices of steel; there was a sigh of relief. Seven months later the finance minister, Dr Manmohan Singh, announced in his famous speech of 25 July measures that liberalized the Indian economy. TISCO, without notice or preparation, was exposed to international competition. B. Muthuraman, the present managing director, who was a part of the senior management then, recalls those difficult times:

'We were thrown into the water—virtually. We didn't know how to swim. We needed to learn to swim as we were in the water. Those were tremendously difficult days with a labour force of 78,000 producing 2 million tonnes of steel. In the first half of 1992-93, TISCO profits were slender. It was close to a loss situation. When Dr Irani took over the company, his first priority was to put the company back on track. That is what he did for 10 years as managing director—from 1991 to 2000. There was tremendous cost control, improvement measures, value engineering programs, improvement initiatives, programs like Total Operating Performance, improving product mix, improving market mix, improving yields in the plants and reducing the labour force from 78,000 to 38,000, which we are today. Those were extremely difficult days. If we had not done that we would not have survived today.'

'The awakening call came from a report of Arthur D. Little (ADL)—management consultants from USA. Basically it was a scary report. It said virtually that your profits have been poor, your steel plant is in bad shape. You have poor equipment, bad technology, high cost and bad marketing processes. Your decision making process is poor. Your IT systems are inadequate. The company was okay with respect to the era that it earlier lived in. No company could have been prepared given the situation of the previous 30–40 years with the government controlling everything—how much steel to make, what type and at what price.'

'With Dr Irani coming in, a sense of great urgency came in. Dr Irani was a fanatically feverish person, if I may use the word, who came to correct the situation. He made us realize the gravity of the crisis. If the company did not do anything urgently, we would have slipped into further trouble in the wake of the reforms process. So Dr Irani said: "We must act yesterday. We must act today morning, instead of afternoon. We must decide today instead of deciding tomorrow."'

'Neither Mr Russi Mody, the previous MD, nor I, if we were put in that situation, would have been able to save the company. Our style of management is different. When the ship is sinking, you have to put a man who is at the steering.'

'It was at that time that I saw Mr Ratan Tata at work. He sounded different from any one of us. He made us question the unquestionable; he made us think big and gave us confidence. He empowered us. He brought team spirit in us. He made several visits to Jamshedpur during those dark days of Tata Steel and encouraged us, enthused us and energized us. Dr Irani had just taken over as Managing Director. Under this leadership and with Mr RNT's guidance, the company was brought back from the brink of disaster. Watching Dr Irani from close quarters was a terrific lesson in managing difficult and tough situations.'

'There are times when you need to behave like a General. There are times when you need to behave like a coach. Times when you need to behave like a gardener who looks at the weeds and water to see that tree is growing. There are different times. It calls for different types of leaders. The '90s were a great time of crisis for the company, brought about by external forces. And that crisis found a fantastic leader in Dr Irani.'

Dr Jamshed J. Irani

Dr Jamshed Irani—known to his close colleagues as 'Doc'—was a JN Tata scholar for his M.Sc. in Metallurgy at Sheffield University when 'Sheffield' was synonymous with steel. He was awarded the Gold Medal by the university for his Ph.D. J.R.D. Tata got a letter from the director of the JN Tata Endowment giving him the news of this young man's achievement. JRD sent the letter back with the remark: 'If ever this young man wants to return to India ask him to first knock on the door of the Tata Iron & Steel Company.'

Dr Jamshed Irani came and met J.R.D. Tata but joined research at British Steel. When he finally returned to India, he was placed in the research department of TISCO. He was not happy there for there was much research to be done and his boss was insecure with a highly qualified person under him. After six months, he decided to take up his lien on his old job at Sheffield. They offered him even better terms than earlier. Senior colleagues in TISCO who saw enormous potential in the young man, including Homi Bodhanwala, managed to get him to meet J.R.D. Tata and Sumant Moolgaokar on a plant floor visit. The young man said he was going back. They realized he was dissatisfied and restless for meaningful work. They moved fast and shifted him first to quality control as chief metallurgist and then to operations. That was just the beginning because Dr Irani soon became superintendent, and then general manager. It was not long before he was president and managing director.

Dr Jamshed Irani gives a lot of credit to his predecessor Russi Mody who was then managing director: 'He was a good team builder and could enthuse people.' When Mody was made head of operations in 1972, he would gather about fourteen or fifteen young people every morning over a hearty breakfast in a designated place. They were all supposed to bring their own breakfast. Some brought a simple breakfast like sandwiches, but Mr Mody would have the most sumptuous plate of all with Parsi delicacies like 'Prawn Patya' for his colleagues to share. They would plan for their work that day. It also built up a great camaraderie. Russi Mody was not a technical man and these discussions helped him gauge people and technical problems. For example, he appointed Dr Irani to be in charge of the steel melting shop although some old timers told him Irani was mainly a research man. That was a wise decision and proved that Mody was an astute judge of people.

Sadly at the time of the greatest crisis in 1992, the essential teamwork required at the top level was missing. Dr J.J. Irani was appointed managing director in July 1992. In 1993, Russi Mody, the chairman, made his exit from the company. On 19 April 1993, Ratan Tata was appointed chairman of Tata Steel. Thereafter there was harmony and cooperation. 'The Chairman (Mr Ratan Tata) was always for the latest technology. He allowed me to manage,' says Dr Irani. Jehangir Ardeshir, who was the principal executive officer to Dr Irani says, 'TISCO has always had the right type of leadership at the right time. Someone up there loves it.'

Today Dr Irani is wistful about those happy days and there is a tinge of regret in Dr Irani's voice as things went awry in the last few years of their working life. But Dr Irani credits Russi Mody as playing a big role in building his career.

Just in time for the crisis, the company found the right man in Dr Irani.

16

The Struggle for Survival

In 1992, after the board appointed Dr J.J. Irani as managing director, he was flying back to Jamshedpur. On that flight, he wrote on a pad:

'What I, as CEO, must do.

1) Develop a personal vision—what do I want to accomplish in life.
2) Tell the truth about current reality.
3) Do the tough things no one else wants to do.
4) Restructure the TOP TEAM, if necessary.
5) Build a powerful guiding coalition—management and board.
6) Guide the creation of a shared VISION.
7) Take the responsibility of being the main change agent.
8) Create endless opportunities for two-way communications.
9) Create opportunities for innovations in the rank and file.

10) Maintain focus.
11) Realign HR systems, overcome obstacles.
12) Model the desired managerial behaviour—above all maintain CREDIBILITY.
13) Preserve the core values of the TATAs (and my own).'

On the technology side, as general manager he had already taken the first steps for the modernization of the plant. He had also exposed his staff to quality management after his 1988 visit to Japan.

He now had to tackle the second and third points from his list which were perhaps the most difficult points on his agenda. Not only would these be unwelcome, they would also be greeted with substantial opposition. He informed the staff and the workers about the serious situation facing the company and he took a grip on the toughest job others would hesitate to undertake. He had to modernize not only the plant, but also the mentality of TISCO workers and its staff that had grown stagnant over the years.

TISCO had agreement with the union that after twenty-five years of service, a man on retirement could recommend his son or a close relative for a job. With that clause, an immense loyalty was built between the families and the company for two or three generations. The problem was that with this clause, the staff strength could never be reduced. Dr Irani had sent TISCO'S union leaders to Japan and Korea to see for themselves how with minimum labour steel could be produced. He had already established credibility with the union. He convinced them that if the company were to be saved, this clause had to be overlooked.

Dr Irani was fortunate that V.G. Gopal, the head of the union, was also a man with vision. He was concerned about the workers welfare and could grasp the gravity of the situation. Dr Irani went around all the workshops at the plant. He himself visited all the mines of TISCO, coal and iron ore mines. Fortunately for him, a fine system of joint consultation had been established for decades. TISCO had pioneered the joint department councils for every department—about fifty of them. At the regular meetings, the management and workers met, a worker presiding once and a manager presiding next time. Every three months, he would give an update to his workers and staff. He told them the truth without fear. Even so, it was a bitter truth and he had to act to fulfil his larger vision of saving the company. At one of the meetings he addressed to the workers, one of them stood up and shouted back in Hindi. 'What about job for my son?' Dr Irani shot back: 'I am not worried about the job for your son. I am worried about your job and my job. If we do not change we will be both out of jobs.'

The message rang loud and clear.

The next step was to work out separation terms, keeping in mind the good relationship that had been built over decades between the company and the workers. The separation terms were that with some exceptions of essential staff and workers, an employee could avail of this scheme and continue to get the last drawn salary and Dearness Allowance (DA) till he/she reached the age of retirement of sixty years. The employee could take a second job simultaneously as well as get the last salary cheque he/she was drawing at TISCO.

On his point of modelling 'desired managerial behaviour' to maintain credibility, he ensured for example that during his entire term of his office, he never bought a car for himself. One from the office pool was being used by him. This may seem a negligible point to many but the workers could conclude that the man's principles and actions were of one piece.

The workers were being trained in giving quality service. One day a worker came and said, 'For eight hours a day, you expect us to pursue quality at the plant. But many of our company homes are leaking.' Without even asking for further details of the number of houses, the extent of repair or the cost, he ordered immediate action by his town division staff. Not a day was lost. And with that the tradition of fairness and trust was left intact.

Each employee, literate or illiterate, was a member of a quality circle. No financial incentive was given to the employees who did better than others. But an emblem was given to each person, giving him a sense of personal pride. And this again was personally done by the managing director. As Tata Steel entered the twenty-first century, every single department had been certified for ISO-9001 systems.

Every year, about 3,500 workers and staff accepted the voluntary separation scheme. The number came down from 78,000 to 48,000 in 2001 by the time Dr Irani stepped down as managing director. It came further down to 38,000 in 2006. 'It was inclusive growth,' says Irani, 'and though the family size had to be reduced, there was no coercion but firmness. I always believe in communication with workers. They realized we had no hidden agenda, no pretences.'

Irani wondered after giving workers and staff the most generous separation terms ever whether his decision was correct. But a few years later his finance people told him that he had made the right decision. Irani laid down the retrenchment policy. Satyanand Pandey, vice-president of human resources and his department, worked out the details of separation in each case. The terms were not only generous but risky and a more timid chairman than Ratan Tata would have tried to rein Irani in. But he did not do so.

At the same time, Dr Irani did not flinch from tough measures when they were needed. He was leading a life and death struggle to save a company. He was not in his position to win a popularity contest. The first year, he asked all heads of departments to cut down their budget by 20 per cent. They all knew that some flab could be shed. To their surprise, he called them next year and asked for another 20 per cent cut. One officer piped up: 'But our budget is already 80 per cent of the previous year and now you want us to cut it by another 20 per cent?' Bluntly he replied: 'I don't want to hear anything,' and that meeting was adjourned.

Such a turnaround had to be a team effort. The production team, headed by Dr T. Mukherjee, was told by Mr Tata, after his first AGM as Chairman, 'I want you to cut down the costs by Rs 500 per tonne and raise the net profit by Rs 1,000 per tonne. Dr Mukherjee said that after 4 months, in December 1993, he had cut down the cost by Rs 300 per tonne and was quite pleased with himself. But Mr Tata reminded him: 'Your assurance of Rs 500 per tonne is with me', indicating he expected further cuts. Dr Mukherjee said, 'He appealed to my pride. In the next 3 months, before the financial year was ended, we had cut down costs by Rs 500 per tonne.'

Dr Irani was able to do all that he did it was because of the support and sensitivity of his family. During those ten years of crisis when TISCO was turned around, his wife Daisy made sure that no domestic matter that would disturb him came to his ears. She also took over management of the social side of TISCO so minor issues would not impinge on her husband's time.

The personnnel department of the company has a confidential check up on the performance of officers and in the case of Dr Irani twenty forms were circulated. No officer's name appeared on the form. Dr Irani scored ten out of ten in decision-making, nine out of ten for competence and efficiency but got only one mark for social ability. Unfazed, Dr Irani replied: 'My wife could score ten out of ten in that.'

B. Muthuraman was in charge of special projects, the most important one being to see through the Cold Rolling Mill (CRM). The speed at which this was put up was probably a world record. The Japanese supplier of equipment said that the fastest time it had been erected was thirty-two months. Irani said he wanted it in twenty-nine months.

TISCO did it in twenty-six months. CRM today contributes 50 per cent of the entire production. When Dr Irani wanted to introduce it, with a capacity of three million tonnes, the board rightly asked, 'Do you expect to sell all that?' He was confident. The product is particularly suitable for the automobile industry which grew beyond the expectations of anybody and every tonne of steel was in great demand.

And the beauty of this recovery of Tata Steel is the teamwork on all fronts 1993 onwards. Workers suggestions were not only invited but often implemented. They rose in their own esteem that their ideas and suggestions mattered to the management and recognition and courage in innovation were appreciated. And so it was an all-inclusive effort.

In July 2001, Muthuraman took over a company which could have become non-existent a decade earlier and Dr Irani's leadership had left it on sound footing for his successor.

Muthuraman's appreciation of the difficult and life-saving work done by his predecessor is apparent when he says: 'I was fortunate that when I came to office, there was a foundation. So you can pause, reflect, see how you can build the building, get an architect and can expand. That has been my role in the last 5-6 years. I want to underline the importance of the period 1992–2000. I have been a lieutenant to Dr Irani during this time. Without a person of his calibre, his forceful personality we could not have made it. Dr Irani is not always a consultative person. I tell him that, "I am different. Your style and my style are different. But our goals are the same."'

Before long a welcome and rather unexpected acknowledgement came from an international source.

17

A New Dawn

The Indian steel industry accounted for 6 per cent of the GNP, 16 per cent of the excise duty collection and a total investment of $19.5 billions as on 2001. It employs 4,51,000 people directly and for each direct worker, there were many others, bringing employees who find a livelihood from steel to 2.1 million people plus 2,50,000 people in the foundry industry.

As stated in the previous chapter, in the short span of little over a decade, TISCO moved from the danger of being an obsolete plant to a world-class steel-maker. A few months before Dr Irani stepped down as managing director in July 2001, Peter F. Marcus, the editor of *World Steel Dynamics* (WSD), came down to Jamshedpur and examined every aspect of Tata Steel. Based on certain criteria and a variety of factors the selection was made. The companies were positioned considering the following factors:

1) Operating costs
2) Ownership of low-cost ore and coal
3) Favourite location for producing RM

4) Skilled and productive workforce
5) Price paid for electricity
6) High quality and niche products
7) Degree of 'pricing power' with large steel buyers
8) Dominant in region
9) Balance sheet
10) Borrowed funds and equity on a favourable basis
11) Management is experienced, aggressive, proactive
12) Low legacy (retired-worker) costs
13) Ongoing cost-cutting efforts
14) Cost position of nearby competitors
15) Owns downstream steel using businesses
16) Domestic market growth rate
17) Proportion of domestic sales

Company	Ranking	Score
Tata Steel	1	131
Usinor (Arcelor), Europe	2	129
Posco, South Korea	3	127
CSN, Brazil	4	123
Baosteel, China	5	121
China Steel, Taiwan	6	119
Gerdau, Brazil	7	116
Nucor, USA	8	116
Car-Tech, USA	9	112
Nippon Steel, Japan	10	111
Severstal, Russia	11	111
Dofasco, Canada	12	109

The report is detailed and thorough, mentioning even the age of each Indian CEO, designation, date of retirement, etc. In a letter to Mr Muthuraman, who was to succeed Dr Irani, Peter Marcus wrote that he was publishing in May his core report on the Indian steel industry: 'I will explain to the readers why your company is India's only world-class steel maker and one of the few steel companies in the world with such a standing. This viewpoint is based on a variety of items including your own raw material supply, low operating costs, a special company culture, good profitability, expansion prospects, and location in a country in which steel demand should grow substantially in the future.'

The words 'a special company culture' are worth noting. In spite of radical and sweeping changes in the previous decade, the culture of TISCO had remained intact.

The journal singled out Tata Steel as a world-class steel-maker because of the following factors:

1) Spirit: A private company since 1907, Tata has a long tradition of providing outstanding benefits to its employees.

2) Reserves of coal and iron ore of its own. Its coal reserves are among the best in the country. It needs to supplement its own coal with about one-third medium-quality foreign coal.

3) Very low operating cost by global standards

4) Unique company-developed processes that help the company to make use of its low-cost raw materials. In the case of coking coal, it stamps down its coking coal just before it is charged, which increases the density per cubic metre from about 750 to 1150 kg. In the case of iron ore, it engages in special practices to reduce the phosphorus content.

5) High profits. According to WSD's analysis, the company is currently generating free cash flow of $50–100 million per annum (about $28 per tonne shipped) despite very depressed conditions in the global steel industry (AD 2001).[1]

6) The generation of record profits in the year ended March 2001

---

[1] When there was a surplus capacity of 25 per cent over the demand.

7) Dedicated management and workers. Tata has never had a strike in 65 years.

8) A new cold-rolling and galvanizing complex that has the potential to add substantially to operating profits—perhaps about $50 million per annum—as it is ramped up in the next two years.

On the other hand, it lists other factors which adversely affected Tata Steel. They were:

1) It is located in Eastern India, whereas much of its market is in western India.

2) Freight costs by land are the highest in the world.

3) The nearest port is 250 km away.

The journal notes that the company is considering building its own port. The journal concludes:

'Overall, we view Tata Steel as an extraordinarily well positioned company' and adds 'we think that long term investors should seriously consider ownership of the common stock.'

It also notes that recently small steel plants in the western region that have come up are 'very efficient with modern machinery and low operating costs'. On the positive side, it says 'government is spending 100 billion dollars on the road systems, over the years'.

Also, automobile units are booming, resulting in great demand for cold-rolled steel. The negative side: Before a project gets a 'go ahead', it may still need 45–50 clearances in India.

It observed that unlike any private steel company, Tata Steel runs an entire city and pays for it, spending $21 million a year (about Rs 1,000 crore).

World Steel Dynamics ranking in the following years for Tata Steel was:

2002 : Rank – 3
2003 : Rank – 4

2004 : Rank – 1
2005 : Rank – 1
2006 : Rank – 1

The ranking of a steel company can fluctuate from year to year, but the important question is not the ranking. What counts is the in-built and intrinsic strength of the company. An alert management needs to review it regularly and build upon it. It also needs a long-term vision to envisage the future.

Part VI

Going Global

There are certain corporations the world around, which stand out from their fellows. They need not be the largest or the most prosperous in their country or even in their given field but their achievements and traditions are epochal and in peoples' minds identify the trade or industry to which they belong with themselves.

They may be in trade or commerce opening up new frontiers and new territories, such as, for instance, the East Asiatic Corporation of Denmark, or they may be established in one place in a basic or key industry.

The Tata Iron and Steel Company is such a corporation. It is part of the geography and landscape of India—as much a part of her as her great mountain ranges and rivers.

J.D. Choksi
Director, Tata Industries, 1948–1968

18

Planning for the Future

SECURING THE BASE

'We have two guiding arrows. One points overseas, where we want to expand markets for existing products. The other points right here, to India, where we want to explore the large mass market that is emerging. Not by following but by breaking new ground in product development and seeing how we can do something that hasn't been done before,' said Mr Ratan Tata. When asked how he expected the group to shape in the future, he replied: 'We will be a group that probably has an equal division of businesses within India and overseas. With a large multinational workforce, our outlook as a group will hopefully be truly international, so that wherever we may establish operations in overseas markets, we would come to be regarded as a local enterprise that merely happens to be owned by an Indian corporation.'

Dr Irani left a solid foundation on which his successor B. Muthuraman could expand at home and abroad.

There are Greenfield projects at Kalinganagar in Orissa,

at Chhattisgarh and at Jharkhand. At each of these Greenfield sites English medium schools (co-education) will be set up. Of these three sites, at the time of writing, Kalinganagar seems well on its way. Tata Steel has been allocated the land and has rehabilitated most of the inhabitants, training young people of the dispossessed families in some of their own technical institutes. The first phase is expected to be completed in 2009-10 with a capacity of 3 million tonnes per annum followed by the second phase of additional 3 million tonnes per annum. The two other projects are more long term. Tata Steel has plans to expand the capacity of its existing Jamshedpur plant to 6.8 million tonnes per annum by 2008 and 10 million tonnes per annum by 2010.

In the 1850s, steel first came into wide commercial use mainly for rails for the railways. It was in 1889 that the Eiffel Tower placed before the world the use of steel as building material. While steel is being used, further use of steel can speed up construction. In construction steels, one of the most advanced companies is Bluescope Steel in Australia with which Tata Steel has a joint venture. It specializes in colour-coated steels and will focus on the preengineered building sector. Four of these plants using their patented know-how will be opened in northern, southern, eastern and western India.

Since 2004, Tata Steel has reinforced its efforts to securing its raw material base. For coking coal, it has a stake in the Australian company, Carborough Downs Projects Company. It also has a joint venture with Sila Eastern Company in Thailand, which is a reliable source of high-quality limestone. An Indian company called Rawmet Ferrous Industries, Orissa, specializing in value-added ferro-chrome, has also been purchased.

Before exploring abroad, Tata Steel has ensured that it has its own ports at Dhamra, Orissa, expected to be operational in 2010. It has a joint venture with the famous Japanese Line NYK in an in-house shipping company to increase operational efficiencies and reduce its costs and keep a strategic control of its logistics. It also has a joint venture with a company specializing in port operations, freight forwarding, clearing and chartering. Tata Steel has moved on an internal consolidation as a springboard for future acquisitions and partnerships with other companies abroad. The platform prepares it for the take-off.

Looking ahead, Titanium is the metal of the future, vital for aerospace, defence and high-technology engineering and the company is making efforts to acquire rights for Titanium mineral mining in Tamil Nadu.

REACHING OUT

The hand of history has woven the tapestry of Tatas and in no company is it more visible than in Tata Steel. Just over a hundred years ago, Jamsetji Tata was requesting the secretary of state for India, Lord George Hamilton, for the cooperation of the British government in starting India's first steel works. On the hundredth anniversary of the registration of the Tata Iron & Steel Company in 1907, this company won the bid to purchase the Anglo-Dutch steel giant CORUS. And so the wheel has turned a full circle.

For a century, steel was usually made either in an integrated facility or in different facilities within the same country. In the emerging global scenario, Tata Steel has embarked on a strategy of de-integrated production of steel. It uses resources where it is most economical. For example, rather than carrying bulky iron ore and coal over long distances, it makes semi-finished products in countries rich in raw materials and then takes it to countries with first-rate finishing facilities and high demand.

Tata Steel's managing director B. Muthuraman had a study made of leading countries and raw material sources in the world and on that basis, the strategy was formulated. For example, Japan and West Europe are good to take technology from but they are not high-growth economies. As the twenty-first century dawned, it is the South-Asian countries that have high growth in steel consumption and so, Tata Steel acquired NatSteel in Singapore and Millennium (now Tata) Steel in Thailand, both very fine companies.

Ferro-chrome is an essential element for the production of stainless steel and carbon steel. Tata Steel's history and initiative have ensured a head-lead in the mining of ferro-chrome. In 1949, its geological department discovered high-quality chromite in Sukinda Valley. Over fifty years, the company has strengthened its position as a key supplier of chrome ore worldwide by becoming one of the largest exporters of chrome ore on a global scale. It is the only Asian company to enjoy 'main producer' status in major steel plants of Japan and Europe. It was not surprising then that when the South African government wanted Tatas to make investments in their country, the Tata's chose to set up a plant to refine chrome for the export market. South Africa has over 45 per cent of the world's Ferro-chrome production and Tata Steel's facility at Richard's Bay is expected to be functional by the end of 2008. It is hard to believe that 15 years earlier Tata Steel was facing the prospect of non-existence. Now it could move again and spread out its wings beyond the shores of home.

South African Deputy President, Ms Phumzile Mlambo-Ngruka unveiled the plaque to commemorate the construction of the ferro-chrome plant of Tata Steel at Richard's Bay.

Tata Steel has some strategic assets worth talking about. Its proximity to raw materials, its modern steel-making processes, its dominant domestic position in a large and growing market and its high-margin-high-value applications make it amongst the most profitable steel companies in the world.

A further asset is that in terms of earnings before interest, depreciation and taxation, it is the highest compared to global giants. The comparative figures in 2006 are: Tata Steel—42 per cent; Posco, Korea—28 per cent and Arcelor-Mittal—16 per cent. With all these assets, the chairman, Ratan Tata, was planning how to lift Tata Steel from a small plant on the world steel front (ranked 58 in terms of steel production) into a big player. It so happened that when the chairman as well as the CEO of the Anglo-Dutch steel plant Corus came to meet Mr Tata in Mumbai, he found that the chemistry of the two companies clicked. Corus had the same philosophy and culture as the Tatas. The Corus team were very impressed by their visit to Jamshedpur and discussions started on Tatas acquiring Corus. The bid for the Anglo-Dutch steel giant, Corus, was first made at the end of 2006. Before the bid of 455 pence per share could be put to the shareholders, CSN, Brazil's steel producer with an annual production of about 5 million tonnes per annum, came into the picture wanting to make a larger bid. The issue then came under the Rule 32.5 of the Takeover code and subsequently, the auction was held.

According to the rules of the auction, there were a maximum of nine bids. The first eight bids were open bids, and if no winner emerged in the first eight rounds, the final bid was to be a sealed bid. All this was on the same day. On the night of 30 January 2007, the chairman of Tata Sons and Tata Steel, Ratan Tata; N.A. Soonawala, vice-chairman of Tata Sons; Krishna Kumar, executive director, Tata Sons; B. Muthuraman, managing director, Tata Steel and its vice president (finance), Koushik Chatterjee, were closeted in a room at the Taj Mahal Hotel. At the London auction room, on Tata's behalf sat Arunkumar R. Gandhi their expert on mergers and acquisitions.

Putting in the sealed bids in London through their representatives both sides had a 50:50 chance of winning. The last round of open bid the eighth—having been 590 pence, the Tatas were to decide what the next bid should be. They were prepared to go higher, but they put the figure of 608 pence per share. Muthuraman says at that point it was not a strategy, nor was there any particular skill involved. 'Why did we put the final number of 608 pence? Who gave us this number? I don't know.' Their opponent, CSN's bid, when opened, was 603 pence. 'If we had bid higher, we would have had to pay more. Who protected us from paying more? Not we.' A man of faith, he ascribes it to a higher power.

19

Part of the
Indian Landscape

The establishment of TISCO unleashed the latent industrial energies of the Indian people. It gave India its base of heavy industry and overcame its pathetic dependence on the West. More importantly, an enterprise of this magnitutde owned and managed entirely by an Indian gave other Indians a sense of confidence, never felt before. The birth of the company stirred into activity the ingenuity of Indians which lay suppressed and unexplored for centuries. The skill of metallurgists that made the Iron Pillar of Delhi and the best sword blades in the world was revived. Many who speak of the great potential of India and its people can only wonder at the passion, dedication, and ingenuity of those that began India's journey to the industrial and economic power it is prophesied to become soon.

TISCO enabled other industries to start, and provided the climate and the material for engineering that gave birth to several companies, the largest of them being Telco (now

Tata Motors). When TISCO started one hundred per cent of India's steel requirements were imported. In 1927, India made 30 per cent of her steel, mainly through TISCO; in 1934 she made 70 per cent of her requirements. TISCO generated confidence of Indians in India, and though intangible, confidence too is an economic asset.

The doyen of economists in the early twentieth century, Sir Alfred Marshall noted: 'I do not think that manufactures are more conducive to prosperity than agriculture is, unless they evoke initiative. A score of Tatas might do more for India than any Government, British or Indigenous, can accomplish.' He obviously had TISCO in mind.

Picture for a moment, India at the turn of this century. It was a vast agricultural nation being governed by an imperialist nation. Not only did Jamsetji Tata dare to give India the sinews of industry but he created around himself men who dared to think along uncharted ways. While the Tatas imported technology at the turn of the century, they did set their own stamp in other fields. The eight-hour working day in 1912, provident fund and leave with pay in 1920, were all innovations way ahead of Britain and the USA. In 1914, Burjorji Padshah made the unusual request to the board of directors that 1.5 per cent of the net profits of the company be put into a fund 'for introducing among working classes at Sakchi improved diet and better clothing'. This does not appear to have been accepted but in 1918 Thakkar Bapa did manage to so drastically reduce the prices of food and clothing—in some cases, like cloth, to two-thirds the price—that every worker did have a chance to be reasonably fed and clothed.

In 1928, the first of a variety of bonuses were introduced—one for production, another for efficiency, etc. In 1937 India's first profit-sharing bonus was introduced for the entire staff. TISCO has maintained its head-on lead in industrial benefits. Union leader V. Gopal proudly said: 'Our facilities system has flowed into the public sector. For example, others used to grant half pay in cases of injury while a man was on duty. Tata Steel, from the very beginning, paid full wages for anyone injured on work plus his compensation. Other steel plants had to follow suit.'

It has learnt through trial and error, followed by genuine care and bold imagination. The machinery of joint consultations pioneered by TISCO is noteworthy, but more so is its genuine participative spirit that has made all employees feel they are treated equally and with respect. It has led to the emergence of a feeling that the company belongs to the men who work in it even though the workers have no directors on the board.

Jawaharlal Nehru said Jamshedpur had 'become symbolic in some ways of the growth of Indian industry'. Over the years the company has been associated with the development of the Indian nation as no other company in India. Subhas Chandra Bose believed that if Indian industry was to survive, this company had to survive. The relevance and importance of this company was realized even by its imperial rulers. In 1924 the British-Indian government acted to save TISCO. The company has paid back its debt to the nation in the years that followed, by handsome contributions to the national exchequer, and the training of thousands of technical personnel, as well as in furnishing to the public sector some of its top managerial personnel, like Russi Billimoria (TISCO's chief of personnel) who became the chairman of the Steel Authority of India Limited (SAIL); Mr Samarapungavaran, a TISCO metallurgist who became a managing director of the Bokaro Steel Plant and later chairman of SAIL; and R.N. Sharma, chairman of Coal (India).

When a non-Tata company is in difficulties, the government may call upon TISCO to put it on its feet. The West Bengal government did so in the case of an ailing company, the National Steel Rolling Mill, and TISCO brought it out of the red.

TISCO has not rested on its oars. It has moved with the times and responded. It recognized that Jamshedpur was fast becoming an island of prosperity in an ocean of poverty that is Bihar. It consciously decided to reach out in a very big way to lift the surrounding areas to a decent standard of living, and today has the largest rural development programme of any company in India more than 600 villages in 2007.

Tata Steel has created not just steel for a nation. It has created within itself and around it, a climate of freedom. Its environment enables independent initiative and offers its staff the facilities of finance and personnel needed to fulfil objectives that have little bearing on its self-interest. An Asian Games medallist says: 'The company does not ask us "What will benefit the company?" It asks us "What will benefit India?"'

A senior officer says, 'Tisco is not just a factory. It is a civilization. Its township is the cleanest and you can safely drink water from the tap. Each department over the years developed its own norms. That is why when it started standardization in the '90s, it was difficult but all were imbued with the same spirit.'

It is only in this dimension that one can understand why one of the largest corporate companies of India spends time in mending broken homes and arranging for local adivasis to buy buffaloes or poultry; why its staff for years with an antique machinery could give over a 100 per cent production; and why not a single case has been to a labour arbitration court for over six decades.

Tata Steel is not so much an organization as an organism that has grown and continues to grow, taking its own shape and form as the years go by. After some research into the company this writer realized that it was not a company he was writing about but the unfolding of a miracle. No doubt the miracle will need to be nursed sensitively every day— and down the line—but it is a miracle nonetheless. Who knows, perhaps through its experiments, its trials, its errors and its triumphs, Tata Steel may be feeling its way to the shaping of an industrial society yet to be born.

Epilogue

The Spirit of Jamshedpur

In his Foreword of the book, Chairman of Tata Steel, Mr Ratan Tata says, 'The strength of the company is embodied in the spirit of people and the unbelievable will to win'. What has created the spirit? What has sustained it? What has made it effective?

It is environment that creates the spirit. The spirit in turn is sustained by a set of values of the management, concern for its staff and workers, transparency in its dealings and sense of fairness. It is these that inject confidence in the people at work.

Mr B. Muthuraman, Managing Director, says: 'Tata Steel is a unique organization. It has a formal structure, just like all organizations have. But structure is not important, in Tata Steel. Its people feel empowered, enthusiastic and energized. They take on initiatives and targets beyond the obvious and beyond what purely a structured approach would result in. Beyond these structures is the passion and aspirations of these people. This is what has brought them to where they are there today'.

The management is not snatching the credit but credit also is due to them. In the most difficult decade of 1990s, there was transparency. More than any technical modernization during Dr Irani's time was his ability to frankly communicate with his staff. Basically, Tata Steel is reaping the dividend of being true to its tradition. The structure is not to be under-estimated, especially of labour consultations and different committees. Being true to its traditions result in a sense of trust. Management and labour then are not two separate entities but are part of the same endeavour. It must not have gone unnoticed that most difficult years of the least profit, Tata Steel slashed all the costs, but not where it effected its work for needy villagers in the surrounding areas.

In 1996 the then prime minister, Mr P.V. Narasimha Rao, invited top industrialists in India and urged them to give 1 per cent of their net profit for promoting social work not concerned with the benefit of their own staff and workers. Mr Ratan Tata and Dr J.J. Irani were invited and when the prime minister said 1 per cent should be given, Mr Tata and Dr Irani looked at each other knowing fully well that Tata Steel had done much more. On return to Jamshedpur Tata Steel's figures were worked out and the results showed that in the years Tata Steel made less profit, the ratio that it had utilized for philanthropic purpose was higher. As the profits improved, the percentage spent on philanthropy came down. In the worst year 1992-93, keeping to Tata's policy, they did not cut down on their philanthropic efforts. The chart given below gives the figures, which according to Dr Irani include, 'All Corporate Social Relief (CSR) activity which do not relate to their own employees and their families and townships (Jamshedpur, Noamundi, Joda, West Bokaro, etc.)':

Year	Profit After Tax (Rs crore)	Expenditure on Social Service (Rs crore)	Percentage
1992–93	127	27	21
1993–94	181	24	13
1994–95	281	31	11
1995–96	566	29	5
1996–97	469	33	7
1997–98	322	39	12
1998–99	282	37	13
1999–00	423	38	9
2000–01	553	38	7
2001–02	205	36	17
2002–03	1012	46	5
2003–04	1746	63	4

Tata Steel's social welfare falls broadly into three sections:

Urban — Community development initiatives (employees also overlap with non-employees).
Rural — Through Tata Steel Rural Development Society
Tribal — Through the Tribal Cultural Society

Each of these bring opportunities for training in literacy, vocational training for youth and women, and accent on health of adolescents and women and the work keeps expanding to domestic management for housewives.

Empowerment and income generation are the main objectives of the social thrust. The spirit of Jamshedpur is such that the desire for uplifting others has permeated the thinking of many in the society and in their own way they do it. For example, an organization called 'Basanti' educates women in hygiene, HIV awareness and child care.

One day a wife of a worker said, 'Our husbands are away for most of the day but we don't know what they do and where they work.' So visits to the steel plants were arranged in small groups of 25 or 30. The women saw the hard work their husbands were doing. One of the workers said, 'Formerly when I went home in the evening my wife used to drown me with all her problems of the day before I could sit down. Now she prepares a cup of tea for me and after I am refreshed she tells me what her day has been like.' It had an impact on the domestic life of the workers and their performance at work.

The Tata Steel Rural Development Society (TSRDS) sought to galvanize entire villages towards a better quality of life. The mechanisms used by the Society include self-help groups, early childhood educational programmes, youth leadership programmes, farm-level training, training of rural artisans and marketing rural handicrafts.

If Jamshedpur was the first in erstwhile Bihar in family planning and in its blood bank, it is because of the spirit of its public not just the support of companies. This writer met a young man of another Tata company in Jamshedpur who made it his mission every Sunday to go to villages at his own expense and organize blood donors.

The remarkable thing about Jamshedpur is not only its willingness to assist in cases any natural disaster but the speed with which the Tata Relief Committee moves into action. It is led by Tata Steel with its expertise, but other companies contribute handsomely in funds and man power. For example, when the Bhuj earthquake took place before agencies in nearby Mumbai could get their act together, the Tata Relief Committee of Jamshedpur was already on the scene. It decided wisely that to restore normalcy the most important thing to do is to start schools again. The moment children go to school the parents feel that normalcy is beginning to come in. Tata Steel set out a programme and executed it with incredible speed sending their own steel from Jamshedpur.

It was the same with the Tsunami disaster. The Tata Relief Committee took up marine relief operations in five districts along the sea coast. It distributed boats with engines as well as nets for fishermen who had lost their livelihood. They also set a water desalination plant for safe drinking water. They constructed over a thousand houses and five larger community-cum-rain shelters which could serve as protection against calamity but also a Panchayat office, a Village Knowledge Centre or a marriage hall.

Tata Steel strengthened its initiatives in various areas of income generation to further the objective of economic value creation and self-reliance of the villages in the vicinity of its operational areas. The primary tools used by the TSRDS to encourage its target groups to earn a better livelihood were: irrigation projects, agri-extension projects, animal husbandry, floriculture, vocational training, etc. Promotion of economically viable industries including rural handicrafts and other cottage industries among the rural people were also taken up by the company.

The youngest of the welfare societies, the Tribal Cultural Society was started in 1993 with the specific purpose of promoting their welfare. Jharkhand—where Tata Steel is based—is predominantly a tribal area, and promoting the welfare of the tribals is the Tribal Cultural Society's mission. Sixty-eight per cent of Jharkhand's potential workforce is either unemployed or underemployed. Its young people have diverse aspirations and even more diverse needs. In its first decade of work literacy has grown over 13 per cent and the 400 promising youngsters who otherwise would have starved of growth have received fellowships worth Rs 25 lakhs each year.

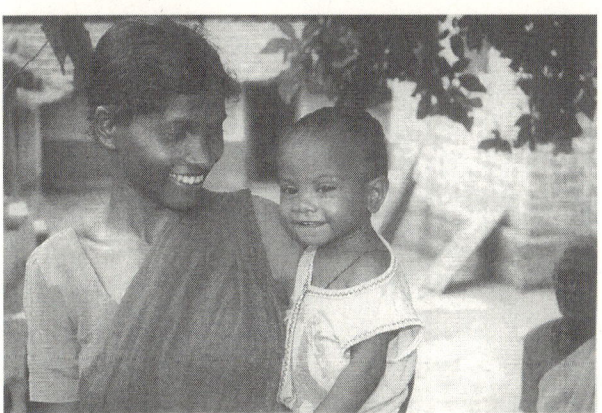

Tribal woman carrying a baby

Ganesh Ram, a non-matric crane operator says: 'I used to wake up out of horrific dreams where I saw my children in tattered clothes struggling for existence (as he probably had to do).' Over the years his five children received the Jyoti Scholarship and his eldest daughter completed her Honours in Mathematics from the prestigious Benaras Hindu University. The youngest one hopes to be a doctor one day.

Satyendra Kumar would have had to discontinue pursuing his Masters in Commerce just then he qualified for a 2-year stipend from the Tribal Society and is now a senior audit officer in a Cooperative Society.

There is one Youth Resource Centre in all the 18 villages that come under the project. Girls and women are given an opportunity to understand physical hygiene and explore livelihood opportunities as per their talent. Adam Porsch of the Bill and Melinda Gates Foundation says: 'I am so very impressed with the new adolescent centre and health clinic. This visit has been the highlight of my time in India.'

Other institutions like the National Foundation for India and Development Alternatives like to associate with a Tatas run society and extend technical and even financial contribution because they know how well managed it is. The Ramakrishna Mission also helps for specific youth programmes.

Both for adolescents and for women vocational training is given the utmost importance and with it, where needed, opportunities for marketing their products be they bead-handicraft, candle making, poultry, mushroom growing, grass-mat weaving or horticulture for ladies. The boys are prepared to enter technical institutions.

When one knows the stories of some women one is moved by their suffering and how the Tribal Society has helped them to come out triumphant. Sangeeta suffered from impaired hearing and speech which deprived her from pursuing her artistic skills. Her physical handicap compelled her parents to pay a handsome dowry to get her married. But marriage only compounded her sufferings. She was tortured by further demands until finally she left with her two children. She was given vocational training. Today she stitches and embroiders clothes and is self-sufficient. After many a summer she has a reason to smile.

Fifty community health providers were given a year long training at the main hospital and the Tinplate hospital and they passed equipped to give first aid, administer drugs and injections and maintain records. Some tribal women participated in computer-aided literacy programmes.

That is not all. At their Annual Prize Distribution Day handicrafts were laid out for visitors and tribal women participated in joyful music and dance with their families giving expression to a latent urge in their hearts which would otherwise have never been fulfilled. Women were rescued from the corridors of illiteracy and despair.

With all this their tribal traditions are retained.

'What you do for one of these, you also do for ME.'

Acknowledgements

This book, unlike others of mine, was written in two stages separated by a span of almost two decades. I greatly enjoyed doing intense research in 1982 and discovered gems of other books for reference. I resumed work on this book on Tata Steel again in 2006 on the eve of the Centenary. The first period of the book deal, mainly with written material on the romance of the early years. The rest of the book, especially the last quarter century is mainly based on interviews with those who have written the history of Tata Steel by their deeds.

The books that appealed to me most were those which gave human insights into those early years, like that of Mrs Lillian Ashby, wife of the Superintendent of Police in Jamshedpur and hostess at the Directors' Bungalow there. Her book *My India* was enormously helpful in providing a view of life in Jamshedpur during that time. John Keenan's book *A Steel Man in India* was invaluable in more ways that one. Keenan rose to be the general manager and a stadium is named in his honour in Jamshedpur. Books that gave hard facts were: Frank Harris' *J.N. Tata: A Chronicle of His Life* and Lovat Fraser's *Iron and Steel in India—A Chapter in the Life of Jamsetji N Tata*. I am obliged to Verrier Elwin's work on Tata Steel published by the company in 1957; but for this timely work many facts then available may have been lost today. Research for my book *For the Love of India—Life and Times of Jamsetji Tata* (2004) also helped.

In 1979-80, when I was looking for material for my book *The Creation of Wealth—Tata Story*, I was pleasantly surprised to find that the only company that seemed to have meticulously kept records was TISCO. Almost two decades later, I discovered from the Archives that there was a letter from J.R.D. Tata to the general manager, Jehangir Ghandy, dated 4 February 1941. Mr Tata wrote, 'I have asked Miss Cursetjee to collect documents regarding important events in the history of the Company and send them on to you for use and preservation. Proper arrangements should be made at Jamshedpur to have these indexed and preserved'. It is thanks to his foresight that Tata Steel records are well preserved. Also since 1932, there has been a record of events in their magazine *TISCO News* which again provides a wealth of material.

Apart from the exhaustive archives which have been an enormous help, in the early 1980s I had the privilege to meet the then oldest surviving staff member of Tata Steel. He had joined, if I am not mistaken, in the 1920s and his name was Ghosh. It is people like these who gave me a feel of the company in the early years.

For specific material, I am obliged to the Tata Central Archives and Tata Steel Archives, the latter specially for the photographs.

R.M.L.

Index

For the Love of India:
The Life and Times of Jamsetji Tata

R.M. Lala

'A mine of information on a man who transformed India and dragged it into the twentieth century'—*Sentinel*

Jamsetji Nusserwanji Tata was born in 1839, and in his lifetime India remained firmly under British rule. Yet the projects he envisioned laid the foundation for the nation's development once it became independent. More extraordinary still, these institutions continue to set the pace for others in their respective areas. For, among his many achievements are the Indian Institute of Science in Bangalore, which has groomed some of the country's best scientists, the Tata Steel plant in Jamshedpur, which marked the country's transition from trading to manufacturing, his pioneering hydro-electric project, and the Taj Mahal hotel in Mumbai, one of the finest in the world.

In these as in other projects he undertook, Jamsetji revealed the unerring instinct of a man who knew what it would take to restore the pride of a subjugated nation and help it prepare for a place among the leading nations of the world once it came into its own.

The scale of the projects required abilities of a high order. In some cases it was sheer perseverance that paid off—as with finding a suitable site for the steel project. In others, such as the Indian Institute of Science, it was his exceptional persuasive skills and patience that finally got him the approval of a reluctant viceroy, Lord Curzon.

In *For the Love of India,* R.M. Lala has drawn upon fresh material from the India Office Library in London and other archives, as also Jamsetji's letters, to portray the man and his age. It is an absorbing account that makes clear how remarkable Jamsetji's achievement truly was, and why, even now, one hundred years after his death, he seems like a man well ahead of the times.

Biography
Rs 295

The Creation of Wealth:
The Tatas from the 19th to the 21st Century

R.M. Lala

'The book stands out like a jewel in more ways than one...A writer of the calibre of R.M. Lala invests the entire narrative with simplicity, dignity and elegance'—*Financial Express*

When Jamsetji Tata started a trading firm in 1868, few could have guessed that he was also opening an important chapter in the making of modern India. Jamsetji saw that the three keys to India's industrial development were steel, hydroelectric power, and technical education and research. A century and a half later, the Tatas can claim with justice to have lived up to the vision of their founder.

But the road to success has never been smooth. Appearing for the first time in this edition is the story of how the Tatas, with Ratan Tata at the helm, have had to grapple with change in the post-1992 era of economic reforms. In a frank epilogue, Ratan Tata talks about the difficulties he faced in implementing change, including resistance from his colleagues.

The Creation of Wealth is R.M. Lala's best-selling account of how the Tatas have been at the forefront in the making of the Indian nation— not just by their phenomenal achievements as industrialists and entrepreneurs but also by their signal contributions in areas like factory reforms, labour and social welfare, medical research, higher education, culture and arts, and rural development.

Business
Rs 295

**Beyond the Last Blue Mountain:
A Life of J.R.D. Tata (1904—1993)**

R.M. Lala

An exhaustive and unforgettable portrait of India's greatest and most respected industrialist

Written with J.R.D. Tata's co-operation, this superb biography tells the J.R.D. story from his birth to 1993, the year in which he died in Switzerland. The book is divided into four parts: Part I deals with the early years, from J.R.D.'s birth in France in 1904 to his accession to the chairmanship of Tatas, India's largest industrial conglomerate, at the age of thirty-four; Part II looks at his forty-six years in Indian aviation (the lasting passion of J.R.D.'s life) which led to the initiation of the Indian aviation industry and its development into one of India's success stories; Part III illuminates his half-century-long stint as the outstanding personality of Indian industry; and Part IV unearths hitherto unknown details about the private man and the public figure, including glimpses of his long friendships with such people as Jawaharlal Nehru, Mahatma Gandhi, Indira Gandhi and his association with celebrities in India and abroad.

Biography
Rs 295